AVENUE21. Connected and Automated Driving:
Prospects for Urban Europe

Mathias Mitteregger · Emilia M. Bruck ·
Aggelos Soteropoulos · Andrea Stickler ·
Martin Berger · Jens S. Dangschat ·
Rudolf Scheuvens · Ian Banerjee

AVENUE21. Connected and Automated Driving: Prospects for Urban Europe

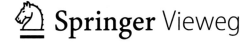

Mathias Mitteregger
future.lab Research Center
TU Wien
Vienna, Austria

Aggelos Soteropoulos
future.lab Research Center and Division of
Transport System Planning
TU Wien
Vienna, Austria

Martin Berger
Division of Transport System Planning
TU Wien
Vienna, Austria

Rudolf Scheuvens
future.lab Research Center and Research
Unit of Local Planning
TU Wien
Vienna, Austria

Emilia M. Bruck
future.lab Research Center and Research Unit of
Local Planning
TU Wien
Vienna, Austria

Andrea Stickler
future.lab Research Center
and Center for Sociology
TU Wien
Vienna, Austria

Jens S. Dangschat
Center for Sociology
TU Wien
Vienna, Austria

Ian Banerjee
future.lab Research Center
TU Wien
Vienna, Austria

Translation & Proofreading:
Maria Slater & Nivene Raafat

Typesetting & Layout:
Jonathan Fetka

ISBN 978-3-662-64142-2   ISBN 978-3-662-64140-8   (eBook)
https://doi.org/10.1007/978-3-662-64140-8

The Daimler und Benz Stiftung is funding the Ladenburger Kolleg AVENUE21. Connected and Automated Driving: Prospects for Urban Europa, that is publishing its findings in this work.

Planung/Lektorat: Markus Braun
This Springer Vieweg imprint is published by the registered company Springer-Verlag GmbH, DE part of Springer Nature.
The registered company address is: Heidelberger Platz 3, 14197 Berlin, Germany

FOREWORD
DAIMLER AND BENZ FOUNDATION

Digital technologies have not only become an integral part of the value chain in agriculture, aerospace, logistics, telecommunications, the media and entertainment, but are now proving to be key drivers of further developments in their own right. In medicine too, we are moving towards data-driven medical science that not only guides medical professionals in diagnosis and in choosing methods of treatment, but is also capable of accompanying patients individually in their course of therapy.

Our transportation systems are on the threshold of a comparable development that is about to have a great impact on our society and will fundamentally transform the face of our cities. Just as the automobile has had a visible influence on the built structure of public space over the past 130 years or so, partially or highly automated vehicles along with closely networked transportation systems for people and goods will emerge as determining factors in the future. With shuttle buses, owned or rented cars, delivery boxes, rail and other public transport, car-sharing services, and rented scooters, we will then have a variety of transport options at our disposal that will need to be orchestrated in real time.

What we are hoping for is plain to see: the prevention of accidents, lost time and traffic jams, along with gains in economic and ecological efficiency, increased safety for the passengers transported and enhanced travel comfort. On the other hand, the infrastructure for these highly complex, interlocking services has yet to be created. We should also bear in mind the far-reaching indirect effects of this revolutionary development as early as possible: comprehensive data on our mobility behaviour will be collected, which must be appropriately used and protected if we do not want to be involuntarily transformed from anonymous transport users into transparent passengers. This mobility transformation could also make numerous occupations redundant, while others could newly arise – which will undeniably require the employees concerned to undergo processes of learning and adaptation that will not always be easy.

Moreover, cities and municipalities will be faced with some questions: who will have to pay for and maintain the necessary technological infrastructure for the traffic of the future? Who will guarantee its correct functioning, even in critical situations? After all, the safety of passengers depends on it. Last but not least, the hetero-

geneous nature of public space and the built structure of existing cities will lead to distortions: while citizens in some urban centres will have a variety of individual transportation options to choose from, in other places these can scarcely be provided to a comparable extent or interact in a practicable way. The needs and expectations of various social groups, urban and rural populations, public institutions, private individuals and companies will considerably diverge.

This book sets out to contribute towards providing a basis for the expected public debate and to provide all discussion partners with the necessary information. Over a period of two years, the Daimler and Benz Foundation funded an interdisciplinary research project for this purpose at the TU Wien in Vienna under the title "AVENUE21 – Autonomous Traffic: Developments in Urban Europe". On the basis of selected cities and metropolitan areas, this project investigated what scenarios can be expected for Europe and what developments are already emerging today on a global scale. In particular, the scientists were at pains to develop scenarios and identify viable social solutions for the future that offer added value for all concerned and can help defuse anticipated conflicts in advance by means of objective classification. May their research findings presented in this volume further contribute to the development of a sustainable understanding of planning that does not react to changes, but anticipates them. Architects, urban planners and citizens alike should thereby succeed in understanding the upcoming changes above all as a historical opportunity to jointly shape the city of the future in a socially attractive, liveable and ecologically sustainable way.

Prof. Dr. Eckard Minx,
Prof. Dr. Lutz H. Gade

Board of Directors of the Daimler and Benz Foundation

FOREWORD
AVENUE21

A reflection on the history of the "European city" (for more on the use of this term, see Chap. 3.2) reveals the close connection between the constructed city, transport and mobility. Each new transport technology has left its mark on the social fabric of the space in terms of the mobility of urban societies and has generated new urban structures. Amongst the "revolutions of reachability" (Schmitz 2001), the car takes on a special role. Unlike trains, trams or aeroplanes, at first no separate transport network was created for private car use; instead, existing public spaces – streets – were simply adapted for car traffic. The resulting land-use conflicts between functional transport on the one hand, and urban liveability on the other were mostly resolved in favour of function, in line with Modernist thinking. Urban spaces were allocated accordingly and their use was laid down in law.

When the impacts of connected and automated transport (CAT) are analysed, it is again assumed that CAT technology will prevail in existing public spaces, i.e. streets. To date, barely any studies have factored in the heterogeneity of road networks or the creative leeway of local planning cultures. In this publication, we counterbalance this reductionism by incorporating a wide range of perspectives. Looking closely at different streetscapes and planning rationalities has relevant implications for any assessment of the technological feasibility and implementation potential of CAT in European cities.

Connected and automated vehicles (CAVs) are just one aspect of a broad technological, economic, ecological and social change: globalization, digitalization, the climate crisis, urbanization, social differentiation and the integration of increasingly diverse cultures are examples of the challenges to which modern societies will have to adapt. Hartmut Rosa speaks of an unending cycle of acceleration when it comes to technology-driven social change that has a significant impact on everyday life (2012, 2013; see Fig. 0.1).

An area where these challenges are particularly relevant is mobility in an increasingly mobile society. For this reason, European cities are on the cusp of a radical change

1 with which the transition to a post-fossil fuel era must be actively encouraged, shaped and realized (Kollosche/Schwedes 2016) and

2 which has gained momentum due to technological innovations in particular and whose potentials and risks are only gradually becoming apparent (Rosa 2013).

This change will neither occur "of its own accord" nor will it be possible to manage it with existing planning practices; rather, it will require collaborative and

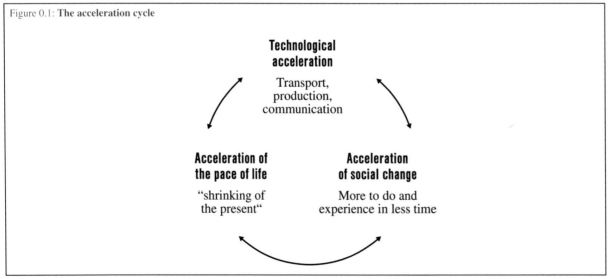

Figure 0.1: **The acceleration cycle**

Technological acceleration

Transport, production, communication

Acceleration of the pace of life

"shrinking of the present"

Acceleration of social change

More to do and experience in less time

Source: AVENUE21 based on Rosa (2012)

proactive learning and action. For this reason, it is vital to investigate the conditions under which CAT can contribute to sustainable mobility and consider the extent to which unpredictable risks may arise.

Underlying our research project is an approach that takes into account the heterogeneity of urban regions and local governance. This method has made it clear that some of the presumed effects of CAT – such as improvements in road safety, the avoidance of traffic jams or a potential change in people's and companies' behaviour – have to be not only reassessed but also evaluated in light of local circumstances. This has significant implications, primarily for near-term planning policy and practice, which will ultimately set the course for future CAT development. To this end, our research project included the identification of fields of action and the development of potential regulatory measures; our findings are presented in this publication.

The questions asked of any changing mobility system are intensified in connected and automated mobility (CAM). As with past transport innovations, the implementation of this technology will have a significant impact on future urban development. The precise nature of this impact is largely uncertain, but it is bound to manifest itself in very diverse ways, especially in smaller areas. The technical developments related to CAT are currently still in the formative phase, as they were throughout the period of our study. This stage is characterized by experimentation and will gradually give rise to solutions. It is therefore also a creative phase: in terms of technological development, but also in terms of policy, planning, administration and civil society.

AVENUE21 Project Team

SUMMARY:
THE IMPACT OF CONNECTED AND AUTOMATED TRANSPORT ON URBAN DEVELOPMENT IN EUROPE

Increasing connectivity and automation of transport is expected to fundamentally reorganize mobility in urban and rural areas and all related economic sectors. Driving assistance systems and connectivity in new cars are quickly becoming established in the entire fleet. At the same time, connectivity and automation are being tested worldwide in the context of demand-oriented mobility services – as yet limited to test areas. Even though the technical feasibility of connected and automated vehicles (CAVs) is assessed more soberly today than during the initial euphoria, this change in mobility systems could have spatial and social consequences as far-reaching as the advent of the automobile some 100 years ago. Preparing for this challenge, better understanding the opportunities and risks of CAVs with regard to more sustainable transport, and hence making CAVs practicable, are therefore the central tasks of spatial and mobility planning at the beginning of the 21st century.

EXPERT OPINION: EXTENSIVE URBAN TRANSFORMATION IS TO BE EXPECTED

Within the framework of the AVENUE21 project, two expert surveys with more than 300 participants were carried out to determine the current level of awareness in urban development and associated fields of the potential impacts of CAVs. The central findings are:

- The respondents have great confidence in the public sector. Policy measures are called for at the local and regional level, but there is no consensus as to the specific nature of these policies.

- The interviewees explicitly desire stronger participation by political actors and civil society in the discourse surrounding CAVs.

- Automated means of transport are expected to have a high potential to displace non-automated means of transport. According to the experts' assessment, the highest pressure in this regard comes from automated car- and ride-sharing services (97.6% of respondents believe that these applications will displace non-automated means of transport). Connected and automated private cars were named second (96.2%).

- The results suggest that a paradigm shift in mobility is imminent. This is shown by the fact that, in the context of automated driving systems, the question "Will this trip require the presence of a human being or will it be delegated to a machine?" will have to be answered first, before addressing the hitherto fundamental question of transport and urban development policy: "Which mode of transport will people choose for a particular trip?"

- The experts expect that this change will not occur evenly across the settlement structure. The earliest and highest suitability for the deployment of CAVs was attributed to streets in industrial and suburban areas, while historical city centres and inner cities are believed to have the worst suitability (see Chap. 3.4).

A SPATIALLY DIFFERENTIATED PERSPECTIVE: THE LONG LEVEL 4

It should be noted that streets simultaneously have to ensure both the functioning and the liveability of towns and cities. The resulting conflicts of use have shaped their development. To date, most studies have largely ignored the fact that with CAVs, the optimization of streets for transportation purposes will undermine their use as public spaces. The decisive role played by policy and planning, whose task it is to find a balance between these two demands, has also been widely disregarded (see Chaps. 4.1, 4.2).

In the case of automated driving systems, this contradiction has an influence on technological feasibility. Their use in a heterogeneous road network, consisting not only of busy inner cities but also motorways, residential, commercial and industrial areas means that no date can be predicted for the impending change (see Chap. 4.4). A gradual process is more likely, extending over several decades, during which CAVs will be deployed only in parts of the road network. During this transition period, conventional means of transport will continue to play an essential but increasingly specialized role. We call this stage the "Long Level 4".

The following findings can be summarized for the Long Level 4 stage:

- For a long time, CAVs will only be available in parts of the city. Previously anticipated impacts – from road safety to traffic performance issues and the potential of reclaiming land currently used for parking spaces – must be reassessed.

- With "automated drivability", the AVENUE21 project team has developed an index that can be used to determine the suitability of road segments for highly automated vehicles. Only publicly accessible data are used for the analysis, meaning that large road networks can be evaluated at little expense (see Chap. 4.4; main finding of Aggelos Soteropoulos' dissertation).

- The Long Level 4 is characterized by a fundamental contradiction: roads that are attractive public spaces constitute the greatest technological challenge and will not allow for automated driving at today's regular speeds. Considerable pressure is to be expected on the development of these streets.

- The uneven deployment of new mobility services in passenger and freight transport will lead to the destabilization of the spatial/transport system from the very moment that automated driving is possible on motorways.

- A differentiated suitability of the road network is likely to have a significant influence on the location choices of businesses and individuals, causing extremely dynamic land use during the Long Level 4.

- Opportunities (better connection to the public transport network, increasing attractiveness of the location for transport-intensive economic sectors, recovery of public space) but also risks (problems of traffic safety, road use conflicts, increasing traffic volume) will be concentrated in peripheral and already car-friendly locations.

- Without any management measures, progressive urban sprawl will be the first spatial effect of CAVs. The increase in accessibility thanks to new mobility services activates land reserves, which will in turn exert pressure on regional land markets.

- The connectivity, automation and electrification of vehicles will lead to considerable municipal fiscal effects that were qualitatively recorded during this project (see Chap. 4.3).

TOWARDS SUSTAINABLE MOBILITY: ADVANCEMENT OF LOCAL AND REGIONAL TRANSPORT POLICY AND PLANNING WITH CAVS

Connected and automated vehicles are part of a present that is characterized by profound social changes (see Chap. 3.1). As in the past, current dynamics of change will be intensified by new transport technologies. They drive change and ultimately determine its spatial effectiveness.

The project team acknowledges that, due to the climate crisis and the frequently polluted roads in European urban regions, change will inevitably be initiated within a paradigm of greater ecological sustainability that also takes social and economic effects into account. The relevance of this aspect was also confirmed in the expert surveys, in which the reduction of environmental pollution, the improvement of traffic safety, the development towards more compact cities and more socially inclusive mobility were named as the most important urban development goals. There is wide agreement that these goals cannot be achieved through new technologies such as CAVs alone. Therefore, urban mobility in particular must be completely reimagined to (1) avoid traffic, (2) shift to more environmentally friendly forms of transport (active mobility – walking and cycling – and public transport), and (3) improve the attractiveness of streets as public spaces.

Implementing this A-S-I strategy – avoid, shift, improve – will require established instruments and policy measures (e.g. traffic avoidance in the 15-minute city) as well as completely new approaches. As CAVs are likely to bring about far-reaching urban transformation, effective measures by dedicated actors are imperative. New actors must be identified and involved, while existing competence boundaries must be questioned.

The following points are an initial framework for a critical reflection on existing planning and policy approaches:

- A discourse analysis of different transport policy levels in the European Union (EU) reveals contradictory objectives. In the EU and in most nation states, CAVs are portrayed as the environmentally friendly, safe, and "smart" modernization of the automobile that is capable of strengthening both business in the EU and inter-European cohesion. This view is contested at the local and regional levels of transport policy where traffic issues are apparent (see Chap. 4.6; main finding of Andrea Stickler's dissertation).

- Full coordination and communication between the various political levels have not yet been achieved as a result of this dissent; opportunities for municipal and regional action have therefore remained largely unexploited.

- Cities determine the everyday living conditions of their inhabitants to a large extent. Local and regional policy and planning therefore offer greater scope and flexibility. In addition, planning and management decisions can be made relatively quickly and be more targeted.

- In the course of a case study on pioneering regions around the world (San Francisco, Greater London, Gothenburg, Tokyo, Singapore), it became clear that in almost all such regions the change in mobility (as a partial aspect of digitalization) is regarded as so fundamental that the administration itself is being restructured. The basic assumption is that the established structures do not have the necessary flexibility to cope with the newly emerging cross-sectional issues (see Chap. 4.5).

- The considerable uncertainties that will arise in urban and mobility planning in the coming years make it necessary to establish reflexive planning and governance concepts in which the possibility of revision is an integrated part of the planning process (see Chap. 4.7; main finding of Emilia Bruck's dissertation).

- Real-world trials, pilot projects, and urban living labs offer great potential to test CAVs as part of a revisable process in line with the objectives of sustainable mobility and to integrate them into the existing traffic system or to develop them to this end.

FIELDS OF ACTION FOR THE NEXT FIVE TO TEN YEARS

The influence of the spatial context on the implementation of CAVs and the fact that a shift towards sustainable transport also requires classic political planning suggests that the transition period can be shaped locally. During this project, narrative scenarios were developed to emphasize such local management opportunities. For the scenario writing aspect, the constellation of actors from the fields of policy and planning and their mindset were selected as the key factor. By involving several focus groups, practical knowledge flowed into the scenario process (see Chap. 5).

The results of the analyses and scenario work were combined in practice-oriented action plans. They examine central questions that cities and urban regions will have to address in the next five to ten years. An important prerequisite is the formulation of clear objectives in the context of urban and neighbourhood development and transport in order to provide sustainable urban mobility for all (see Chap. 6).

The transition to sustainable mobility requires decisive action. Connected and automated vehicles present opportunities and risks and pose new challenges for urban regional administrations and planners. Requirements and framework conditions must be defined for governance and planning processes (adaptive, controlling, restrictive and/or promoting) so that CAVs can contribute to achieving the ambitious goals of sustainable mobility. The dynamics that result from an uneven spatial deployment of CAVs during the Long Level 4 make it necessary for decision makers to act quickly. Cities and regions have to accept their responsibility to shape the future and cannot wait until issues are fixed at higher policy levels and/or until the new technologies have already been deployed.

Since 2013, cities and regions within the EU have been encouraged to develop and jointly evaluate Sustainable Urban Mobility Plans (SUMPs). In late 2019, a revised version of the SUMPs was presented that considers the challenges – both opportunities and risks – of CAVs. The results of AVENUE21 were partly incorporated into this revised version (cf. Backhaus et al. 2019). In order to ensure the goal of sustainable development at the local and regional level, the main objective is to strengthen the planning authorities and policymakers at these levels, and to support them in their strategies.

OUTLOOK

The Daimler and Benz Foundation has decided to support this research project for another year as part of the Ladenburg Research Cluster. During the first two years of research, the importance of rural areas in the context of connected and automated mobility became apparent. Furthermore, due to the minimal number of studies on spatial planning, or approaches to management and planning, that focus on Europe, international authors were invited to contribute to a reader on the opportunities and risks of CAVs in the context of spatial and mobility planning in urban and rural areas. This publication was published in early 2021.

ACKNOWLEDGEMENTS & PROJECT CONTEXT

The study presented in this book is the product of a research concept developed at the TU Wien in 2016 and initiated by the Daimler and Benz Foundation. Its specification benefited greatly from dialogue with representatives of the foundation. It is thanks to the foundation's foresight that this project was supported as a Ladenburg Research Cluster and generously funded with a budget of €880,000.

Throughout the entire duration of the project, we were always able to benefit from energizing technical discussions with an ever-growing network of colleagues. We would like to use this opportunity to express our heartfelt thanks to each and every one of them. The constant dialogue with AustriaTech, and above all Martin Russ and Christian Steger-Vonmetz, was of vital importance throughout the project.

Over the past two and a half years, the project team has taken part in numerous workshops, conferences and presentations, which both bolstered our communication with specialists and enabled a deeper understanding of technological developments. Furthermore, the project team made targeted efforts to encourage discourse between researchers and practicians in a field that was only just emerging when our project began.

MEANS OF COMMUNICATION

In June 2017, a two-day review meeting was held in Vienna in order to discuss with international colleagues the initial focus points that we had identified for our research. The exhibition *Hello, Robot* at the MAK – Museum of Applied Arts provided us with a fitting backdrop.

During the largest expert survey to be conducted to date in the extended urban development sector in Europe, we managed to contact over 300 individuals from the fields of research, administration and planning in October 2017 and autumn 2018.

Following an invitation to the Urban Future Global Conference in Vienna, the initial results of the scenario process were presented to a large audience in March 2018 and, subsequently, improved on the basis of the input and suggestions from these international actors.

In April 2018, the AVENUE21 team organized various focus groups at the TU Wien in which attitudes and scope for action were critically discussed.

During the project, we not only sought dialogue with specialists: in May 2018, we compiled our intermediate results for our contribution to the Lange Nacht der Forschung. This context allowed us to have many inspiring discussions with interested visitors. We would like to thank the, in part very young, audience for opening our eyes to brand-new perspectives on connected and automated vehicles.

At the suggestion of AustriaTech and BMVIT, the Austrian Federal Ministry of Transport, Innovation and Technology, the initiative was taken for an urban dialogue on connected and automated mobility, which the project team was able to help shape from the outset. Then, in the course of revising the action plan "Automatisiertes Fahren: Automatisiert – Vernetzt – Mobil" ("Automated Driving: Automated – Connected – Mobile"), this initiative was formally committed to paper and its implementation launched.

WE WOULD LIKE TO THANK ...

Stefan Arbeithuber	MO.Point, Vienna
Helmut Augustin	City of Vienna
Gerald Babel-Sutter	Urban Future Global Conference
Martina Baum	University of Stuttgart
Jerome Becker	TU Wien, Vienna
Patrick Bonato	Graphic design & illustration, Innsbruck
Robert Braun	IHS, Vienna
Johann Bröthaler	TU Wien, Vienna
Kris Carter	City of Boston
Francesco Ciari	Joanneum Research, Graz
Fabian Dorner	TU Wien, Vienna
Linda Dörrzapf	TU Wien, Vienna
Angelus Eisinger	RZU, Zurich
Tomoyuki Furutani	Keio University, Tokyo
Arnulf Grübler	IIASA, Laxenburg
Susanna Hauptmann	Kapsch, Vienna
Philipp Haydn	mobyome, Vienna
Wencke Hertzsch	City of Vienna
Andreas Käfer	TRAFFIX, Vienna
Sven Kesselring	Nürtingen-Geislingen University
Christoph Kirchberger	aspern.mobil.LAB, Vienna
Wolfram Klar	AustriaTech, Vienna
Hans Kramar	TU Wien, Vienna
Daniela Krautsack	Mykamabook
Nico Larco	University of Oregon
Chris Leck	Futures Division Ministry of Transport, Singapore
Thomas Madreiter	Urban Development and Planning, City of Vienna
Katharina Manderscheid	Universität Hamburg
Anna Mayerthaler	Wiener Stadtwerke
Alexandra Millonig	AIT, Vienna
Teresa Morandini	TU Wien, Vienna
Michael Nikowitz	BMVIT, Vienna
Graham Parkhurst	UWE Bristol
Paul Pfaffenbichler	BOKU, Vienna
Patrick Poh	Land Transport Authority, Singapore
Karl Rehrl	Salzburg Research
Jack Robbins	FXCollaborative Architects, New York City
Peter Rojko	Thinkport VIENNA
Martin Russ	AustriaTech, Vienna
Katja Schechtner	OECD, Paris
Claus Seibt	University of Kassel
Eriketti Servou	mobil.LAB, Technical University of Munich
Vanessa Sodl	TU Wien, Vienna
Henriette Spyra	BMVIT, Vienna
Christian Steger-Vonmetz	AustriaTech, Vienna
Karin Tausz	SBB Swiss Federal Railways, Bern
Gregory Telepak	City of Vienna
Arjan van Timmeren	AMS Institute, Amsterdam
Karen Vancluysen	POLIS, Brussels
Marlene Wagner	buildCollective, Vienna
Angelika Winkler	City of Vienna
Cornelia Zankl	Salzburg Research
Renate Zuckerstätter-Semela	SUM-Nord, Vienna
Jakob Zwirchmaier	TTTech, Vienna

as well as the participants of the "Städtedialog Automatisierte Mobilität" in Bern and the "Lange Nacht der Forschung" in Vienna in 2019.

LIST OF ABBREVIATIONS

ADS	Automated driving system
ADUS	Automated driving for universal services
AI	Artificial intelligence
AR	Augmented reality
B2B	Business-to-business
B2C	Business-to-consumer
BASt	Federal Highway Research Institute, Germany
BEIS	Department for Business, Energy & Industrial Strategy, UK
BEV	Battery electric vehicle
C2C	Consumer-to-consumer
CA	connected and automated
CARTS	Committee on Autonomous Road Transport for Singapore
CAV	connected and automated vehicle
CAM	connected and automated mobility
CAT	connected and automated transport
CCAV	Centre for Connected and Autonomous Vehicles
CIAM	Congrès Internationaux d'Architecture Moderne
C-ITS	Cooperative intelligent transport systems
DAB	Digital audio broadcasting
DDoS	Distributed denial of service
DfT	Department for Transport, UK
DMB	Digital multimedia broadcasting
G2C	Government-to-citizen
IBA	International Building Exhibitions
ICT	Information and communication technology
IoT	Internet of Things
LSEV	Low-speed electric vehicle
MaaS	Mobility as a Service

MPT	Motorized private transport
MLP	Multilevel perspective
NGO	Non-governmental organization
ODD	Operational Design Domain(s)
PHEV	Plug-in hybrid electric vehicle
PPP	Public-private partnership
PPPP	Public-private-people partnership
PT	Public transport
SECAV	Shared electric connected automated vehicle
SFMTA	San Francisco Municipal Transportation Agency
SIP	Cross-Ministerial Strategic Innovation Promotion Program
SIP-ADUS	SIP-Automated Driving for Universal Services
SUMP(s)	Sustainable Urban Mobility Plan(s)

CONTENTS

CONNECTED AND AUTOMATED TRANSPORT

THE ROAD AHEAD

1

© The Author(s) 2022
M. Mitteregger et al., *AVENUE21. Connected and Automated
Driving: Prospects for Urban Europe*,
https://doi.org/10.1007/978-3-662-64140-8_1

1.

CONNECTED AND AUTOMATED TRANSPORT: THE ROAD AHEAD

At least since the turn of the millennium, it has become clear that Europe's cities are facing a rising number of increasingly demanding challenges. Climate change and global technological development are two considerable global drivers that are also governing social change in Europe. These two significant aspects require us to re-examine fundamental questions concerning urban development. They are also set to shape the development pathway of an urban mobility model on the verge of radical change. On the one hand, thanks to automation, we are seeing the development of multiple sensors, and driving assistance and propulsion systems; on the other, digitalization is giving us mobility solutions (MaaS – Mobility as a Service) – controlled via platforms or apps – that cater to and amplify specific demands, thus bringing forth new types of mobility. In a modern, mobile society, transport systems play a crucial role in shaping both objective and subjective assessments of quality of life, and they are inseparably linked to urban development challenges. New technologies and a transformation of mobility based on new mobility types thus also require urban development, urban policy and urban planning actors to rethink their concepts, strategies, measures, processes and instruments.

New mobility technologies offer a whole host of opportunities: digitalization can be an additional tool for urban policymakers and planners, but its potential might only be tapped over time (Giffinger et al. 2018); connected and automated vehicles (CAVs) form a key aspect of smart city strategy design; and comprehensive infrastructural investments in the expansion of digital networks (e.g. 5G and G5 technology), smart curbs and lane or traffic light sensors (Mitteregger et al. 2019) are currently the subject of debate. However, as the ongoing discussion surrounding the concept of the smart city and its existing implementation shows, predictions regarding the impacts of these technologies vary hugely (Hajer 2014, Kitchin 2015, Bauriedl/Strüver 2018, Libbe 2018). In terms of the overall discourse, the need for policy and planning decisions, and problems associated with implementation, certain parallelisms to the somewhat more recent reflections on CAM are emerging.

If assumptions of just how encompassing this looming transformation could be are correct, and the impact of this shift comes close to emulating that of another milestone – the introduction of the automobile – then it is clear that traffic planners will not be the only ones whose minds will be occupied by CAM. The integration of CAVs in the European city (for more on the use of the term "European city", see Chapter 3.2) should therefore be used to reflect on the suitability of and familiar approaches to existing instruments (transport, infrastructures and platforms, as well as management approaches to policy and planning). For this reason, it is crucial to explore and examine – today, not tomorrow – the extent to which new technologies can contribute to the existing goals of sustainable mobility and do so without triggering any unwelcome side effects.

THE OPPORTUNITIES AND RISKS OF CONNECTED AND AUTOMATED TRANSPORT

Connected and automated vehicles are currently the topic of widespread discussion in the media. It is a debate outweighed by reports of technological progress, potential applications and the expectations linked to these innovations in the field of mobility. These articles are accompanied by images of futuristic-looking vehicles and diagrams of connected vehicle networks within a smart city. Even academic publications are dominated by reports on optimized vehicle technologies and the new potential resulting from vehicle-to-vehicle, smartphone and smart home connectivity.

Critical narratives are less common, primarily concern questions regarding the ethics, accountability or authorization of such technological innovations, as well as automated driving systems, and express the population's widespread scepticism. But various scenarios have also given rise to reservations among academics, who cast numerous doubts on the problem-solving potential of CAM. In fact, concerns are being raised that CAVs may not only generate new problems but exacerbate existing ones.

The scientific debate surrounding the future and the impacts CAT will have on the road and settlement structure is dominated by the following points.

- The discourse surrounding CAVs has long been characterized by (successive or disruptive) technological feasibility and/or economic efficiency (Freudendahl-Pedersen et al. 2019). Only recently have there been more frequent considerations of the possible interplay between economic, ecological and social implications. Thus the spectrum of disciplines that focus on the topic has been expanded (Meyer/Beiker 2014, 2016, 2018, 2019).

- The overwhelming majority of studies examining the effects of CAT focus on transport-related issues (e.g. a greater efficiency of or burden on the transport system, choice of transport) and largely exclude questions concerning planning, policy and society (Milakis et al. 2017, Soteropoulos et al. 2018a).

- "Self-driving" vehicles are often portrayed as the "solution" to all the existing negative impacts and side effects of (urban) mobility: they will supposedly help avoid congestion, reduce accident figures to almost zero, consume less energy thanks to "smart" traffic management and thereby reduce the amount of harmful emissions. Moreover, the technology will allegedly grant independence of movement to persons with reduced mobility and thus make social integration (once again) possible (BMVIT 2016b, 2018, Dangschat 2018, BMVI 2018, POLIS 2018).

- While CAVs are being intensively tested in North American and Asian towns and suburbs, where initial trial operations have been launched (Lee 2018), in European cities, the technology is only being tested on selected routes and at low speeds (8–15 km/h; Boersma et al. 2018, Rehrl/ Zankl 2018). Until now, the varied, densely used and rapidly changing streetscapes of Europe's cities mean the situation is still too complex to allow CAVs to be tested.

- The discourse within society and the media and scientific considerations of CAT predominantly refer to the development of vehicle technologies and their connectivity. This discussion is dominated by various engineering disciplines and vehicle manufacturers, as well as globally active companies in the IT industry and business consultancies (Milakis et al. 2017).

- In the media and as part of efforts to market the smart city and "smart mobility", vehicles often appear to be futuristic, luxurious, stylish and elegant. In animated videos, vehicles either glide through a "sanitized" city, which has very few inhabitants and shows no sign of "urban challenges", or they are placed within an expansive outdoor setting, but with the natural world and the environment playing no role other than that of an aesthetic backdrop (Manderscheid 2018).

- The majority of scenarios almost exclusively refer to fully automated driving (SAE Level 5) where, for instance, various levels of market penetration are discussed (usually 10%, 50% and 90%; Soteropoulos et al. 2018a). In addition, these future development scenarios are largely painted from a perspective based on the interests of different actors on the supply side (e.g. the automotive industry, IT sector, network operators; Beiker 2015).

- In addition, models of how streets, junctions and motorways might look in the future are designed with different vehicles being allocated separate lanes, which in reality would result in urban roads that in some places are up to 60–80 metres wide (NACTO 2017). The impacts this design will have beyond traffic itself, such as how pleasant it will be to use the road in question or an increased separation effect, are largely ignored (Mitteregger 2019, Riggs et al. 2019).

- Particularly in advertising, but also within everyday discourses and academic publications, future mobility developments centre around the car (either owned privately or used as part of sharing concepts; Canzler/Knie 2016). Against the backdrop of a generally growing criticism of automobility (and of propulsion systems in particular), these visions of the future aim to craft a new image of the "self-driving" car as a smarter "third space" (alongside people's homes and workplaces).

DEFINING THE FOCUS OF THIS STUDY

The subject of this study is the various potential applications of CAVs on roads in individual and public transport as well as in the various hybrid forms of sharing. This is a systematic study, i.e. it includes all other currently known modes of transportation. Applications for freight transport and, above all, in "last mile" urban logistics are considered. Other potential new transport technologies that are expected to play a role in the transport system of the future (such as drones) are explicitly left out of this study.

This investigation analyses the impacts of CAVs on the European city, both regarding the normative framework, which suggests a specific approach to policy and planning and the analytical framework, which comprises criteria regarding urban spaces, sustainable mobility and heterogenous urban society.

We will be examining the most imminent phase of the transition. The areas of action (see Chapter 6.3) concern guidelines that are to be introduced in urban develop-

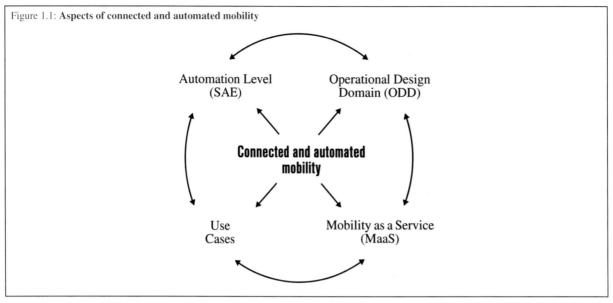

Figure 1.1: **Aspects of connected and automated mobility**

ment within the next five to ten years. It is understood that CAVs will not be able to be implemented uniformly beyond the settlement structure during this phase due to the heterogeneity of roads and their use. Subsequently, phenomena of a social and spatially selective implementation of CAT will be the focus of our observations. This requires a spatially and socially nuanced analysis, and this publication represents the initial steps towards this aim. At the same time, an extensive demand for research is becoming apparent.

Indeed, the overall debate concerning CAM is – as previously mentioned – primarily focused on technological progress (which assistance systems will be available and when?) and the futuristic design of vehicles, especially cars. The assistance systems are divided into groups based on how responsibility for the driving of the vehicle is shared within the human–machine interface. Internationally, these categories are based on the SAE system (SAE International, formerly the Society of Automotive Engineers), the SAE standard J3016 or the BASt (Federal Highway Research Institute) system in Germany. Most previous studies, and above all publications in the media, are limited to these classifications, and in terms of the technological possibilities, concentrate on fully automated driving (SAE Level 5). In this report, however, we will focus on highly automated driving, i.e. Level 4 (SAE 2019, see Chapter 4).

The question concerning when automated vehicles will be authorized is also being discussed in public as well as among researchers. We are thus dealing with an international race being conducted on two fronts: one between companies in the classic automotive and IT sectors to develop the technology, and one between nation states or their subnational authorities (federal states and cities) to authorize such technologies on their roads.

The authorization of various levels of automation primarily depends on their area of application (ODD – Operational Design Domain) such as motorways, rural areas, inner-city districts and industrial estates, as well as further environmental factors, such as the time of day or the weather. Given the myriad complexities of traffic conditions as well as the availability of high-speed internet connections, the decision to authorize the use of automated vehicles over the course of time will primarily depend on how certain situations regarding traffic can be organized in a safe way in mixed traffic environments.

Climate change and congested road networks, however, mean that there is also a need for urban regions in Europe to revolutionize their transport systems. Not only does this require new, post-fossil-fuel forms of transport, but new mobility concepts and behavioural changes. A growing number of digital platforms are appearing that allow various vehicles to be reserved and used for a variety of journeys, the costs to be calculated and for users to exert a certain level of control: MaaS. Car and ride-sharing services are generally considered to play a powerful role in this process. The aim of this transport revolution is thus not to swap current vehicles for "more intelligent" models, but, above all, to completely redesign urban mobility by (1) avoiding travel altogether, (2) shifting to more environmentally friendly forms of transport (active mobility – walking and cycling – and public transport) and (3) improving the quality of public space and transport (A-S-I strategy). Such a shift would drastically change the face of the European city.

As only certain elements of the aspects mentioned here (level of automation, ODD, use cases and MaaS) are discussed or considered in scenarios – and tend to be examined in parallel, not in relation to each other – it seems relevant to always examine these four aspects as a whole (see Figure 1.1).

As the AVENUE21 project has not been designed to quantify future mobility developments in European cities, the aim will not be to compile an "impact assessment" in the classic sense. Rather, the purpose is to outline and attest to the opportunities that are available to a broad circle of actors in urban and mobility planning to play an active role in shaping the upcoming transition.

It is not only the far-reaching consequences of CAM but, ultimately, also the widespread scepticism about the effects of CAT, notably among large parts of the population, (critical) social scientists and especially among transport policy planners at the local level, that make it necessary to discuss the terms under which CAVs are implemented sooner rather than later. The aim should be to avoid any negative impacts on sustainable urban development or for certain subspaces and social groups. It is thus the objective of the AVENUE21 project to develop alternative scenarios for actors in local/regional policy and planning management that clearly point to immediate needs for action but also highlight the possible impacts of various proposed activities.

APPROACH AND KEY AREAS OF FOCUS

THE NEAR-TERM IMPACTS OF CONNECTED AND AUTOMATED VEHICLES ON THE EUROPEAN CITY

2

© The Author(s) 2022
M. Mitteregger et al., *AVENUE21. Connected and Automated
Driving: Prospects for Urban Europe*,
https://doi.org/10.1007/978-3-662-64140-8_2

2.1

AIM OF THIS STUDY

The aim of the *AVENUE21. Connected and Automated Driving: Prospects for Urban Europe* research project is to examine the near-term impacts CAT will have on the European city and to analyse the potentials – both positive and negative – of this new technology. By shifting the perspective away from the technological possibilities and their likely applications and towards the possible effects on planning policy, the impacts in terms of urban development and urban societies, we are able to focus our attention on an area that has until now largely been given little, or, in many cases, only selective, consideration. As the many discussions among researchers, but also particularly among "urban stakeholders", which took place over the course of the project have shown, this shift in perspective is considered vital by those currently concerned with the acceptance of these new technological systems and their regulated implementation.

The project's objective was thus to depart from the "beaten track" as much as possible and instead seek to identify the challenges and opportunities that would arise from the embedding of future mobility within widespread technological evolution as part of a sweeping process of digitalization. Here the aim is not to make assumptions about how CAT would impact future urban mobility if it reached a market penetration of x per cent, but rather, to consider:

- which policy measures would be relevant to achieve the current objectives in urban and regional development, particularly in light of the urgent need for a fresh approach to transport;

- which urban development challenges will arise from the adjustment or enabling of new forms of micromobility with regard to the redesign of streets, new multimodal transport hubs as well as modified districts and building structures;

- how the social innovations and dynamics of change associated with the predicted processes can be designed (car-free multimodality, new business models, new forms of the "sharing economy", etc.);

- which cooperations and skills will be required in planning and management to achieve this.

These chosen areas of focus were ultimately influenced by the members of the research team at TU Wien, which comprises spatial and transport planners, architects, architectural theorists and sociologists.

Another element of the AVENUE21 project is the ambition to develop a practical foundation for the outlining of key areas of focus that should be addressed by actors in urban and mobility development over the next five to ten years. Here the aim is to play a part in ensuring decision makers in cities and regions take account of the challenges, opportunities and risks of CAT that may arise over the next 20 to 30 years and do so in good time, locate these developments within a broader shift that is reshaping the general framework, consider the vital issues concerning the future and are able to develop specific measures so that, ultimately, they can actively shape the transition to a new mobility system.

Such an approach also includes considering other possible forms that the mobility of the future might take and asking questions that technology alone cannot answer. Fundamentally "thinking in alternatives" (Minx/Böhlke 2006) will be decisive in enabling those involved to effectively take up the challenges of the imminent transition period.

2.2
STUDY DESIGN

Figure 2.2.1: **Structure of the AVENUE21 research report**

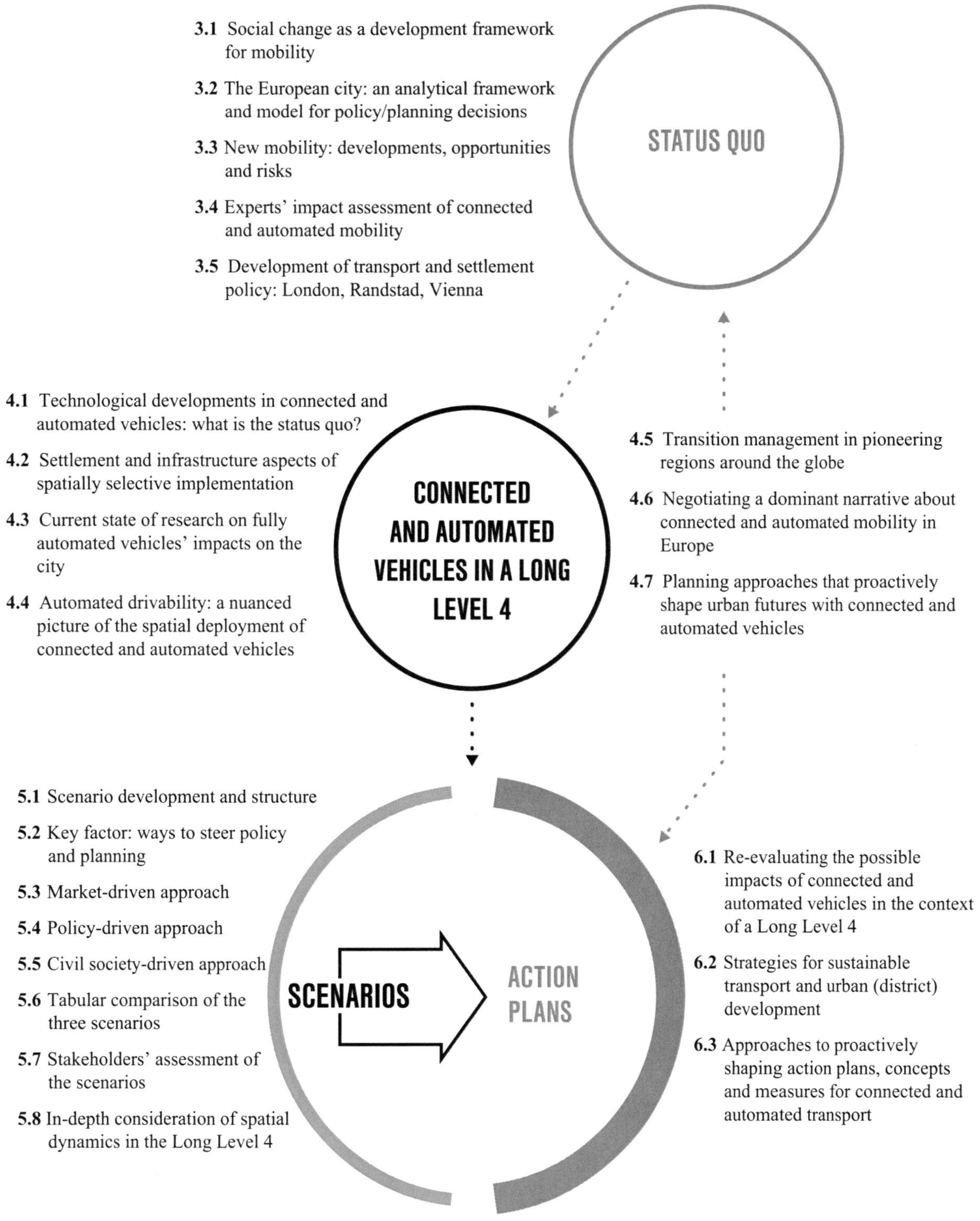

3.1 Social change as a development framework for mobility

3.2 The European city: an analytical framework and model for policy/planning decisions

3.3 New mobility: developments, opportunities and risks

3.4 Experts' impact assessment of connected and automated mobility

3.5 Development of transport and settlement policy: London, Randstad, Vienna

STATUS QUO

4.1 Technological developments in connected and automated vehicles: what is the status quo?

4.2 Settlement and infrastructure aspects of spatially selective implementation

4.3 Current state of research on fully automated vehicles' impacts on the city

4.4 Automated drivability: a nuanced picture of the spatial deployment of connected and automated vehicles

CONNECTED AND AUTOMATED VEHICLES IN A LONG LEVEL 4

4.5 Transition management in pioneering regions around the globe

4.6 Negotiating a dominant narrative about connected and automated mobility in Europe

4.7 Planning approaches that proactively shape urban futures with connected and automated vehicles

5.1 Scenario development and structure

5.2 Key factor: ways to steer policy and planning

5.3 Market-driven approach

5.4 Policy-driven approach

5.5 Civil society-driven approach

5.6 Tabular comparison of the three scenarios

5.7 Stakeholders' assessment of the scenarios

5.8 In-depth consideration of spatial dynamics in the Long Level 4

SCENARIOS → **ACTION PLANS**

6.1 Re-evaluating the possible impacts of connected and automated vehicles in the context of a Long Level 4

6.2 Strategies for sustainable transport and urban (district) development

6.3 Approaches to proactively shaping action plans, concepts and measures for connected and automated transport

Source: AVENUE21

The varied and profound changes taking place and the challenges that already exist in European cities served as a frame of reference for the observation of the possible future outcomes for mobility brought about through CAM. The central question of how the authorization of CAVs would impact the cities of Europe and how local and regional policymakers and planning authorities should respond to these effects will be approached in three stages.

The first involves analysing the existing situation (Chap. 3). First, we outline the context of a comprehensive social change that encompasses technology, the economy, the environment, government and society (Chap. 3.1). Once we have established the "big picture", we place the term "European city" and its trajectory within a historical context (Chap. 3.2) before exploring the shift towards a New Mobility (Chap. 3.3). This is then followed by the assessments of around 300 specialists on the future developments and the relevance of CAT in the context of urban development (Chap. 3.4). The subsequent section provides an overview of the current situation in three European metropolitan regions: Greater London, the Dutch conurbation of Randstad and the Vienna–Lower Austria region (Chap. 3.5).

Based on an assumed prolonged transition phase in Level 4, which we refer to here as a "Long Level 4",[1] Chapter 4 will focus on the technological development of transport and provide a detailed look at the social and spatial implications: the technological developments of connectivity and automation (Chap. 4.1); the temporality of change and the significance of road infrastructure (Chap. 4.2); the current state of research on CAM and its impact on traffic and space, which has been developed as part of a systematic review of international simulations (Chap. 4.3); a critical analysis identifying areas where CAVs could drive within existing city structures, taking "automated drivability" in Vienna as an example (Chap. 4.4). In contrast to the majority of studies, the AVENUE21 project team chose not to concentrate on the "final technological stage" of highly and (in particular) fully automated driving (SAE Level 5), but instead to turn their attention to the long transitional phase to Level 4, during which investments in the future will be made, but the economic and time-saving advantages can only be marginally exploited.

Chapter 4.5 looks at pioneering CAM regions around the globe, particularly with regard to the objectives CAV implementation aims to achieve. This allows us to show which supporting policy and planning management measures, as well as interventions resulting from institutional change, are being implemented. Chapter 4.6 explores how a CAM "narrative" that focuses on the technological possibilities and the associated expectations is being negotiated at various political levels within Europe. The chapter closes with a consideration of the possibilities of reflexive planning, which will be relevant within the context of real experiments in urban districts and given the many questions that remain unanswered.

In Chapter 5, we combine the aspects of European urban development taken up at the outset with the perspectives on settlement development, transport policy and planning during the transition period outlined in Chapter 4. This is done in three scenarios. The scope for action available in policy and planning and the underlying position were chosen as the key factors for scenarios that reflect the standard narrative in order to emphasize various objectives, developments and opportunities to shape mobility and the subsequently derived rationales for action (Chap. 5.2). Here we differentiate between a market-, policy- and civil society-driven approach. To paint a clear picture of the possible future scenarios for mobility and to aid communication in coordination processes with external parties, we have provided a descriptive summary (Chapters 5.3 to 5.5) with figures illustrating each scenario as well as a table comparing all three scenarios (Chap. 5.6).

The subsequent sections then present an evaluation of these scenarios that takes place on two levels: first, the potentials, dangers and possible compensatory actions are examined in focus groups comprising external specialists (Chap. 5.7); second, we take an in-depth look at the social-spatial dynamics of CAT in the Long Level 4 (Chap. 5.8).

Chapter 6 outlines action plans in a deliberate effort to supplement existing policy papers on CAT. Following a re-evaluation of the possible effects of CAVs on the European city (Chap. 6.1), we summarize current discussions on policy and planning strategy papers at various levels (Chap. 6.2). This is taken as a basis to develop seven action plans that are expected to be adopted by urban development authorities and incorporated into their duties and responsibilities (Chap. 6.3).

---

1   The term "Long level 4" is used to express the idea that over the course of a decade-long transition period, CAVs will be shown to be capable of only a small number of driving tasks and that this subsequently means that certain areas of cities will remain off limits to this technology. This concept has been expressed both in current literature on the subject (Beiker 2018, Shladover 2018) and in public statements made by industry (Krafcik in Marx 2018).

2.3

RESEARCH APPROACH AND METHODOLOGY

Due to the wide array of research topics and the diverse work packages, an interdisciplinary approach was chosen that allows the different areas of focus to be addressed effectively (OECD 2015, Kollosche/Schwedes 2016). At the same time, a broad mix of methods was applied, although researchers predominantly used qualitative methods adopted from empirical social research. A large amount of time was taken up by "desk research" on the extensive and rapidly growing field of CAT publications, which was expanded to include other relevant areas, particularly with the inclusion of three dissertations.[1] Additional methods were also used during networking, analysis of (partial) results and the dissemination stage, including expert interviews,[2] group discussions and focus groups.

A significant amount of time and energy went into the creation of three scenarios during the "scenario writings" (see Chap. 5). Here too we turned to our existing network of research colleagues as well as figures active in policy and planning authorities, above all to comprehensively ensure the internal consistency and plausibility of our main assumptions. Moreover, this study's (preliminary) findings were "put to the test" in a series of international conferences and workshops, and, especially as the project continued, incorporated into discussions with the help of lectures and published in a wide range of formats.

Finally, a great deal of importance was attached to the creation of graphics to represent the "future scenarios". Especially given the fact that the possible outcomes of CAT remain abstract and intangible, but that countless scenarios are being talked about, featured in the media and presented in the form of graphics, visual aids are key to initiating discussions. However, it should be mentioned that the AVENUE21 research team is aware that visual representations are always influenced by "subjective" images, moral concepts and beliefs.

In his PhD thesis, Aggelos Soteropoulos examines the potential transport and spatial effects of the implementation of connected and automated vehicles in cities, particularly with regard to the prevailing heterogeneity of urban roads, especially their specific transport and urban planning characteristics. In his research, special attention is given to the investigation of different roads and their suitability for connected and automated transport. The main point of departure here are the current problems, as indicated in test reports, associated with connected and automated vehicles. By taking these underlying conditions into consideration, the author is able to illustrate possible effects as well as potential measures (Chap. 4.4).

2   A list of experts who were directly involved in discussions can be found in the acknowledgements section. Furthermore, over 300 specialists were included as part of an expert survey. It was also possible to discuss ideas with a number of international colleagues at different conferences.

---

1   Andrea Stickler examines the political discourse on connected and automated mobility in Europe and assesses how it interacts with local political and social conditions (Chap. 4.6).

Emilia M. Bruck's dissertation looks at the planning approaches taken for the design of spatial and societal transformation processes in the face of new mobility technologies, such as connected and automated vehicles. In her theoretical examination of reflexive and explorative planning methods, and also as part of a comparative case study analysis, she analyses the extent to which the "appropriate" use of CAVs in cities requires a rethink of local planning approaches (Chap. 4.7).

STATUS QUO

HOW THE SHIFT TO NEW MOBILITY IS CHANGING THE EUROPEAN CITY

3

© The Author(s) 2022
M. Mitteregger et al., *AVENUE21. Connected and Automated
Driving: Prospects for Urban Europe*,
https://doi.org/10.1007/978-3-662-64140-8_3

Figure 3.1.1: The spectrum of developments: trends and focal points

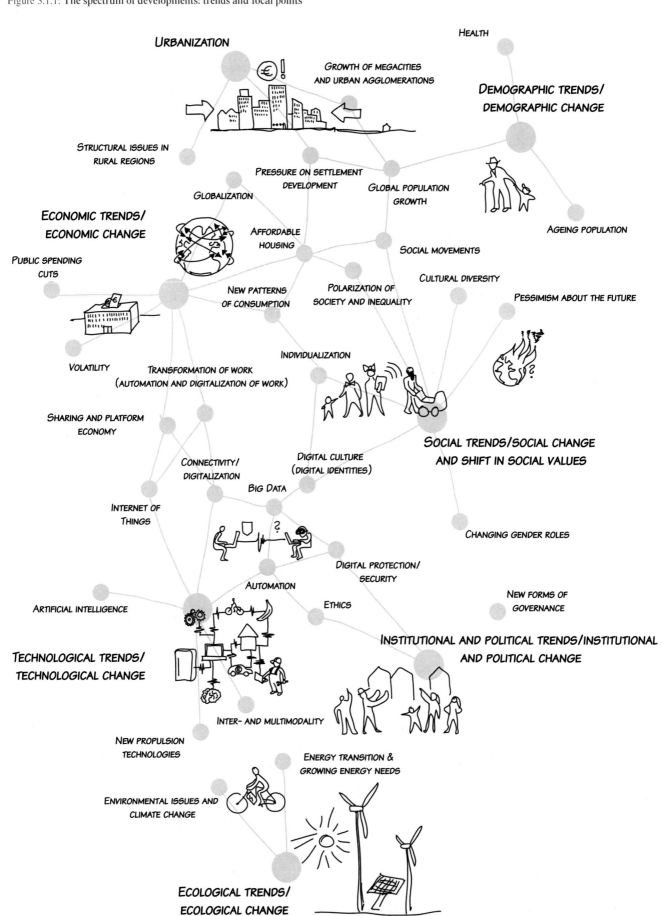

Illustration: Alexander Diem

3.1

SOCIAL CHANGE AS A DEVELOPMENT FRAMEWORK FOR MOBILITY

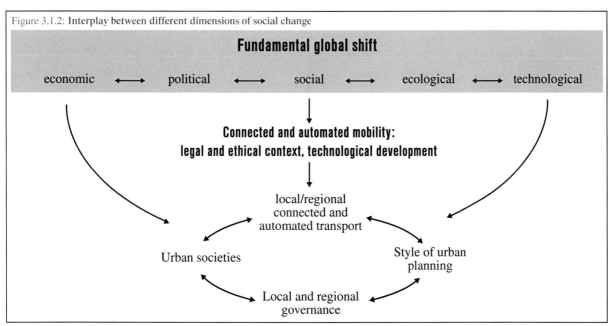

Figure 3.1.2: Interplay between different dimensions of social change

Fundamental global shift

economic ⟷ political ⟷ social ⟷ ecological ⟷ technological

Connected and automated mobility:
legal and ethical context, technological development

local/regional
connected and
automated transport

Urban societies

Style of urban
planning

Local and regional
governance

Source: AVENUE21

Modern society is in the midst of a rapid, intense and extensive transformation. One of the main reasons for this change is the increasing pace of globalization, which is notably responsible for intensifying trade relations, driving capital markets and thus economic competition between nation states, but also facilitating cultural exchange. A key element that makes it all possible is the ability to communicate globally via the Web 2.0 (see Chap. 3.1.1). The second driver of the social transformation we are currently witnessing is the increasing digitalization/digital transformation taking place within the wider context of a large-scale and multifaceted technological revolution. This change not only affects current and future job markets, it also facilitates and promotes additional forms of technological change and has a significant impact on day-to-day practices (see Chap. 3.1.2).

While these two aspects of social change are driven by the economy and technology, the ongoing ecological transformation mostly stems from the consequences of (1) the intensive exploitation of raw materials on a global scale as well as (2) damage to the environment at a local level, both factors in which transport plays a key role (see Chap. 3.1.3). The phenomenon of urbanization – the growth of towns and cities – represents a fourth significant shift (WBGU 2016): even if demographic changes resulting from population growth are mainly impacting Asia and Africa, the shifting population dynamics towards the agglomeration of large cities make these phenomena relevant to Europe, too (see Chap. 3.1.4).

The fifth aspect of social change examined here concerns the new way in which policy and planning decisions are made in a process that includes additional actors ("governance"; see Chap. 3.1.5). Lastly, we examine how the discussed trends are impacting (European) societies socio-economically, socio-demographically, socio-culturally and socio-spatially (Dangschat 2019; see Chap. 3.1.6).

3.1.1 GLOBALIZATION

Globalization is by no means a new phenomenon. Indeed, some analysts claim its origins can be traced all the way back to the global trade relationships that existed in the Greek and Roman empires, or during the Hanseatic League (Jeute 2017). Since the end of the 1960s, however, the term "globalization" has come to mean the renewed intensification of trade relationships, the opening up of capital markets (e.g. with the signing of the Bretton Woods Agreement), the dismantling of import restrictions (tariffs, industrial standards, restrictions on direct foreign investment), the expansion of international air transport (Open Skies Treaty) and, above all, the development and expansion of the internet. The rise of Japan and South Korea, as well as the development of numerous emerging markets in Latin America and Asia, and, more recently, the decision by (ex-)communist states to open up their economies, have led to the intensification of competition between manufacturing regions and of trade relations within the triad of Europe, East and South-East Asia and North America, which has also resulted in goods being produced in different parts of the world and a shift in economic power structures from those countries considered part of the "First World" to emerging markets (the BRICs and Asian Tigers; Ohmae 1985, Beck 1997).

The advent of the internet, and in particular the interactivity that came with Web 2.0, has not only enabled global real-time communication; increasingly we are also seeing the transfer of information between computers and with digitally connected devices using defined algorithms (e.g. on financial markets, in trade relations and, currently, even manufacturing). The proliferation of manufacturing and trade relationships has brought nation states and companies together in the throes of competition between different economic orders, welfare state models, approaches to policy, moral concepts and everyday practices.

For the automotive industry, the development of CAVs and their implementation, this global competition has major ramifications (Porter/Heppelmann 2014). In addition to the growing competition between car manufacturers due to new players entering the market (first Japan, then Korea, China and India), parts suppliers (Bosch, Continental), media businesses (Samsung) and IT companies (Waymo, IBM, NVIDIA, Aurora), as well as mobility service providers (Uber, Lyft), are now increasingly also becoming involved in the manufacturing process for "next-generation" vehicles (Bormann et al. 2018). Moreover, the major export markets are now in fast-growing economies, which have different business models, government regulations and demands.

Public authorities control developments at various levels of transport policy, e.g. by regulating business models (such as via partnerships with foreign investors) and also by funding research. Further regulatory options can have a direct or indirect effect on the future of CAVs, for instance the setting of emission limits or the stipulation of requirements that need to be met for the authorization of highly and fully automated vehicles (see Chap. 3.1.4). The populations of different countries around the globe also differ when it comes to their basic affinity for new technology and their acceptance of highly and fully automated vehicles (Ernst & Young 2013, Eimler/Geisler 2015, Fraedrich/Lenz 2015a, b, Detecton Consulting 2016, Fraedrich et al. 2016, Deloitte Development 2017a, b).

3.1.2 DIGITAL TRANSFORMATION AND TECHNOLOGICAL CHANGE

The term "digitalization" actually means the conversion of analogue measurement and control parameters into discrete (staggered) values so that they can be processed by a computer. However, in common parlance, "digitalization" is understood as the introduction and increased use of digital transmission technology in the economy, in public life and in everyday activities. "Digital transformation", "digital revolution" and the "Fourth Industrial Revolution" are also used to describe a process that has been gaining widespread momentum (Giffinger et al. 2018). Within this context, we are seeing not only more but new kinds of data and data processing technologies (Big Data).

As part of the broader phenomenon of digital transformation, there is plenty of discussion on Industry 4.0, the Internet of Things (IoT), artificial intelligence (AI) and augmented reality (AR). Industry 4.0 is understood as the extensive digitalization of industrial production, with systems connected using state-of-the-art information and communication technology; human-machine interfaces are being redefined. The technical foundation for these developments is provided by intelligent and digitally connected systems, which are to be implemented to make largely autonomous production possible: in Industry 4.0, humans, machines, facilities, logistics and products communicate and cooperate directly with each other (Bauernhansl et al. 2014).

The IoT involves a number of vastly different end devices being linked up and connected via the web – in addition to laptops and smartphones, this includes household devices, household technology (Smart Home), wearable devices and, in the future, CAVs, which former German transport minister Alexander Dobrindt dubbed a "third space" (alongside the home and the workplace) in 2017. The IoT is primarily driven by enhancements and new developments in the field of information and communication technology (Chui et al. 2010).

The IoT is also a major factor when it comes to connected driving as such technologies allow personalized on-trip data to be generated and subsequently capitalized upon. The data, which are mainly generated through connected driving, play a part in managing traffic flows in a way that makes them safer and more efficient, and enable the provision of more effective mobility services (see Chap. 3.3). Data-based business models are also made possible. These data could also have a regulatory effect, for example if the use of certain transport routes and public spaces are priced based on the respective level of use (by the public sector) or based on current demand (by the mobility provider; POLIS 2018: 5).

The significance of artificial intelligence within this context is considerable: if management and control mechanisms for CAT are to be implemented efficiently and effectively, self-learning computer networks and end devices are going to be crucial. Humans could soon be communicating and cooperating with artificially intelligent machines on our roads too. However, questions are increasingly being raised concerning technology-driven surveillance, personal freedoms and the processing of data collected in public space, i.e. on roads (Boeglin 2015, Mitteregger 2019).

Augmented reality (AR) is understood as the use of computer-based technology to enhance our perception of the real world. This information is able to engage all the human senses. However, AR is often understood solely as the visualization of information, i.e. enhancing images or videos with additional computer-generated information or virtual objects that are either superimposed or projected. Some innovative vehicle models already feature this technology: being able to replace the real world with a virtual one while driving is one of the selling points. In addition to the wide range of applications in gaming, AR can also be used, for example, in discussions on future urban development options, which include the presentation, design and management of prospective CAT together with the appropriate traffic infrastructure (Car Trottle 2017).

Alongside technological developments taking place as part of the digital transformation, other developments happening in the field of storage and sensor technology are making it possible for CAVs to adequately and effectively perceive information, process it in real time and decide how to drive (Soteropoulos et al. 2019).

3.1.3 ECOLOGICAL CHANGE

This consists, on the one hand, of climate change, which is most apparent in the warming of the Earth's atmosphere and the subsequent consequences this will have on the sea levels, air and water currents and thus ultimately the weather (drought, heavy rain and flooding, mudflows, thawing permafrost and higher temperatures, particularly in urban areas). On the other, ecological change comprises the extensive exploitation of natural (and, above all, non-renewable) resources (WBGU 2016). These impacts are the result of human civilization, its pursuit of growth, economic systems and unsustainable lifestyles (Brundtland 1987: 1).

With the aim of keeping global temperature rise to below 2°C, just under 200 countries signed an agreement at the 21st UN Climate Change Conference held in Paris in 2015 to limit their emissions of harmful greenhouse gases (particularly carbon dioxide – CO_2 – and nitrogen oxides – NO_x). At a subsequent conference held in Katowice in 2018, delegates agreed to uniform standards

to measure and compare national and regional developments. However, most countries had failed to abide by previous thresholds, with road transport in particular responsible for a continued rise in greenhouse gas emissions (EEA 2017).

In spite of all the technological progress, efforts to make the necessary cuts to emissions have thus far failed, especially in the transport sector: between 1990 and 2014, Germany's emissions dropped from 1,248 to 902 million tonnes of CO_2 equivalent (-25.4%), whereas the transport industry has managed to reduce its CO_2 equivalent by just 1.9% over the same period (from 163 to 160 million tonnes of CO_2 equivalent; BMUB 2016: 8). The target of reducing the industry's emissions by 40–42% by 2030 will allegedly be achieved – according to the "Climate action and mobility" policy embedded in the German government's Climate Action Plan (BMUB 2016: 49–56) – through the promotion of alternative drive systems, public transport, rail travel, cycling and walking, i.e. through an alternative modal split, but also via a digitalization strategy and an increased share of "clean energy". However, the strategy fails to mention how the necessary changes, not only in terms of policy and administrative approach, but also the required behavioural shift among the population, can be achieved.

The reasons behind the transport sector's inability to reach emission targets are lock-in and rebound effects resulting from a widespread dependency on cars. And although engine efficiency is improving, this progress is being effectively cancelled out by ever-larger, heavier vehicles with an ever-increasing engine capacity, the very vehicles that are experiencing growing demand: in 2017, 15.2% of Germany's newly registered vehicles were SUVs, an increase of 22.5% compared to the previous year (Federal Motor Transport Authority 2018). Furthermore, on average, drivers are travelling longer distances and driving at higher speeds (for Austria, see Tomschy et al. 2016: 97). Lastly, there is simply a lack of political will to implement the regulations necessary to fully phase out combustion engines running on fossil fuels (Canzler 2015).

Battery-powered electronic vehicles, as well as the automation and connectivity processes, require greater amounts of increasingly scarce natural resources (e.g. silicon, cobalt, rare earths) that are frequently mined under appalling conditions. What is more, extracting and recycling these materials entails a considerable environmental cost. The electronic vehicle modules in which they are used are also often manufactured under dire, unacceptable working conditions.

3.1.4 URBANIZATION

The demographic and economic developments taking place in emerging economies are also being accompanied by a high level of urbanization, particularly in Asia and Africa. Although, in 2007, the news that the share of the global population living in urban areas had reached 50% was met with enthusiasm by the UN (UN 2008), and despite declarations from the OECD (2015) and the German Advisory Council on Global Change (WBGU 2011) that we are currently living in "the metropolitan century", the truth is that, in addition to spectacular skylines and technological innovations, urbanization brings many challenges, such as increasing socio-economic polarization, a widening divide between urban and rural areas, the loss of traditional values, higher energy consumption and greater emissions.

In Europe, urbanization is not so much a quantitative but a qualitative process. The continent may be on course to see its already impressive level of urbanization (74%) grow further (UN 2018), but this is tending to result in a shift from small towns to large cities and from rural areas to urban agglomerations.

This transformation goes hand in hand with, on the one hand, issues of how to manage infrastructure under- and overload, and on the other, a sharp rise in living costs within inner cities (especially private but also commercial rent costs), which leads to households occupied by lower- and middle-income families being pushed out to the economic and regional periphery (i.e. gentrification). Urban living also brings about a more rapid change in values and a growing cultural diversity that can lead to certain residents feeling overwhelmed (Dangschat 2015a). The urban lifestyles that accompany such shifts are also largely unsustainable, despite all the hopes currently being pinned on the growth of car sharing in major cities (Gossen 2012).

It is thought that such large urban areas have the necessary conditions to help productively shape the mobility transformation. This is what the A-S-I strategy stands for:

- *avoid*: travel as little as possible and avoid forms of travel that harm the environment

- *shift*: generally moving from a mobility model centred on motorized transport and towards a multimodality that promotes other options (e.g. ecomobility through the use of public transport, cycling and walking)

- *improve*: enhancing a wide range of aspects as part of a general improvement of public space and thus quality of life.

The (often wide) range of for-profit free-floating car-sharing services springing up in large cities is often seen as part of this transition, and in recent years these businesses have contributed to a drop in car registration figures in cities and a slight decrease in the number of private cars on the road. However, the number of (short)

journeys has increased (VCÖ 2017). Car sharing is thus in direct competition with active mobility or even public transport. A car-sharing system based on CAVs would mean even greater expectations for profitability (i.e. more providers), lower costs and increased convenience; in other words, it would lead to more spontaneous journeys and unnecessary travel.

3.1.5 FROM GOVERNMENT TO GOVERNANCE

Public administrations first faced criticism for their lack of efficiency back in the 1990s. At the time, the private sector ramped up the pressure to establish an approach called New Public Management, which aimed to replace the bureaucratic, centralized and hierarchical management style with a target-driven, transparent and decentralized model. Instead of imposing rigid rules, managers should increasingly focus on the result, which was to be achieved through cost accounting (instead of governmental accounting), product focus, performance comparisons and contract management. "Responsibility centres" and "flat hierarchies" were introduced with the aim of driving internal and external competition, and increasing individual responsibility (Jann et al. 2006).

Criticism has not only come from the business community: since the 1990s, the general public have shown increasing frustration with the way state services are run. Many felt their voices were not being heard, especially when it came to local and city-wide planning. This resulted in a growing demand for participation and "co-creation" (Sinning 2008), a process also frequently referred to as "the shift from government to governance" (Heeg/Rosol 2007: 504; Bröchler/Lauth 2014).

In political science discussions, but also debates in the fields of organizational sociology and business management, the term "governance" is often also used to signify a departure from structures that are primarily centred on imperative supervision ("command and control"). Rather, drawing on elements that focus on individual responsibility, the managed organizations, units or actors should take an active role in tackling the tasks and/or challenges at hand.

In addition, the term "governance" also includes models of cooperation involving multiple actors. In political contexts, the concept has also come into use both in addition to and as a substitute for "government" and expresses the idea that within the respective entity, management and controlling activities should not just be carried out by the state as the "first sector", but also by the "second sector" (i.e. the market) and the "third sector" (i.e. non-profit organizations, voluntary associations, special interest groups; Heeg/Rosol 2007: 504; Hamedinger 2013: 62). Private sector and civil society actors are thus recognized as resources and instruments that work alongside local policymakers and administrators to intervene where necessary. Consequently, a shift is occurring not only regarding the specific actors involved in such decisions, but also in terms of their authority, responsibilities, competencies and their ability to exercise power. As the number of actors and interests involved in policy and planning decision processes continues to increase, so too does the complexity of institutions and control structures, and thus the need for communication and coordination. Given the rising diversity of actors, and their competing interests, as well as the growing complexity of processes, the future direction of policy and planning management will be key.

This is why, when developing scenarios (see Chap. 5.2), our research team ideally aimed to centre their case studies on three different types of policy and planning management. During this process, the three key sectors – the market, state actors and the non-profit sector – were each focused on separately. Based on the respective characteristics of the market, state and the non-profit sector, our team examine how governance and power relations are shifting in policy and planning management processes and thus how they could have a major impact on the use of CAVs.

It remains unclear how "politics" will approach the challenges of CAT. The experts, however, are in agreement about the importance of engaging with the subject and potential challenges as early as possible (Fagnant/Kockelman 2015), not least to (largely) avoid returning to an urban-planning model based on cars (Jones 2017). On the one hand, it is expected that the EU and the majority of nation states will set different priorities with regard to policy and planning decisions compared to regional and local actors. While the former place greater emphasis on competition (Kauffmann/Rosenfeld 2012), local and regional areas are where the consequences of transport policy play out, and these issues will thus have a bigger role in local decision makers' planning and further policy choices.

3.1.6   SOCIAL CHANGE

As we considered aspects of urbanization, we touched upon certain elements of social change, in particular changes in values and shifting lifestyles and mobility choices. Social change is understood to encompass three aspects:

- *Socio-economic change*
  This concept mainly refers to inequalities in income, which have started to grow again markedly in recent years, and in wealth (Bach 2013, Castells-Quintana et al. 2015). It also refers to shifting social policies that differ both internationally and within nation states, together with labour market risk that varies from region to region.

- *Socio-demographic change*
  This term has long been understood to refer to the growing number of childless and small households (*Versingelung* or increase in one-person households; Hradil 1995) and the ageing of modern societies resulting from growing life expectancy and lower fertility rates (Wehrhahn 2016). In recent years, issues such as migration, refugees and integration have become relevant too.

- *Sociocultural change*
  This concept is largely understood as a change in values (as well as a pluralization of values), a shift away from traditional ties (individualization) and an increased re-embedding in communities of shared values (social milieus) expressed in an increased proliferation of different lifestyles (Dangschat 2014).

Due to different preferences and constraints influencing a choice of residential location, time spent in public space, mobility, etc., these categories bring about largely heterogeneous patterns of distribution (segregation) and behaviours. It can thus be assumed that socio-spatial restructuring constitutes a fourth dimension of social change. This final aspect is particularly significant as many statements on the introduction and acceptance of CAT are made at the national level and without reference to any particular spatial dimension (transport networks, settlement structures, supply and demand profiles, availability).

With regard to a shift in values, the above-mentioned aspects of social change are not uniform, but in fact tend to be polarizing, with new social battle lines being drawn ("Distinction"; Bourdieu 1987) and existing divisions (xenophobia) exacerbated. While younger generations, who are usually better qualified and more tech-savvy, tend to feel positive about the future, a growing section of society is more insecure and anxious. This is particularly true for those groups who are the "modernization losers", having missed the opportunity to partake in economic development ("lift effect"; Beck 2013) or having not been able to keep step with evolving values. The middle classes, in particular, and now gradually the elite too are affected by these new anxieties (Zweck et al. 2015).

Moreover, the advent of Web 2.0 has significantly changed our communication habits, how we manage our time day to day and, ultimately, completely reshaped and revolutionized our relationships with one another. This example uniquely illustrates how technological developments can cut both ways. Smartphones and tablets might well be necessary for the development of new business models as well as make it possible for us to establish social connections with "others" in a pragmatic way (by exchanging or sharing information, time and basic commodities) and allow social innovation. For example, it is thanks to Web 2.0 that the sharing economy was able to get off the ground; crowdfunding and the application of collective intelligence would also be impossible without this technology (Dangschat 2015b).

Yet, on the other hand, the internet also allows "fake news" to be spread at lightning speed, abusive comments to be hurled from anonymous users, our democracies to be undermined (*Disruptive Democracy*; Bloom/Sancino 2019) as well as myriad forms of cybercrime and hacking. Moreover, "smart algorithms" allow discussions and democratic elections to be manipulated through the deployment of social bots.

Some parts of the population in Germany, and particularly in Austria, are expressing serious misgivings about this technology, leading not least to widespread scepticism concerning CAM (Fraedrich/Lenz 2015a, b). This trend can also be seen in the high level of importance (concerning both positive and negative trends) experts attributed to data during the survey we conducted (see Chap. 3.4).

As briefly outlined above, the manifold and, in part, fundamental social change currently taking place will significantly shape the future of mobility. There are some persistent elements that operate as lock-in effects, innovative elements that amplify existing rebound effects or create new ones, and disruptive elements that are exacerbating the anxieties currently felt by swathes of the population. Given the existing landscape, the question of whether and how CAM could help not only solve the current problems of (urban) mobility but also strengthen social cohesion remains, for now at least, uncertain (see Chap. 4.3). In any case, the challenge posed by rapid technological development taking place against the backdrop of social change means that there is a pressing need for policy and planning authorities, along with businesses and civil society, to face up to these issues. It is also important for them to consider how much cities and towns wish to simply adapt to CAVs or CAM or whether they should only permit those business and mobility models that will help achieve objectives outlined as part of sustainable transport and urban development plans (Rupprecht et al. 2018).

3.2

THE EUROPEAN CITY: AN ANALYTICAL FRAMEWORK AND MODEL FOR POLICY/PLANNING DECISIONS

When we use the term "European city" in our research, we do so in full knowledge that the concept, and its associated attributes, have also been the subject of criticism, especially in recent years (Rietdorf 2001, Hassenpflug 2002, Häußermann 2005, Kazepov 2005, Brake 2011, Siebel 2015). On the one hand, a growing connectivity via social media and the globalization of commodity chains, financial transactions and (urban) cultures, as well as increasing regional and transnational links between social milieus, are leading some to question whether it is still possible to make a clear distinction between "city" and "rural space" (Saunders 1987). On the other hand – ultimately due to urban growth in emerging economies – the 21st century has been declared the "century of the cities" by the German Advisory Council on Global Change (WBGU 2016, The Urban Task Force 2003, Läpple 2005, Dangschat 2010).

Beyond this, the image of the European city is one of an "urbanity" that is characterized by functional and architectural diversity, overall cohesion, planned public space, and endless experiences and encounters. Scholars active in the cultural and social sciences, as well as the humanities, however, see the city as a place that is home to a bourgeois way of life, self-organizing, a division of labour and social diversity, tolerance, (civilized) otherness and distance.

Max Weber (1921) developed the concept of the "occidental city" to set the concept of the European city apart from the "oriental city". Weber considered the city to be an economic and social hub that was shaped by the market. Simmel (1903) considered the city to be home to the money economy and an increasingly rational way of living. Today the European city is either contrasted with the "American city" (Bagnasco/Le Galès 2000, Kaelble 2001, Le Galès 2002, Giersig 2005, Häußermann/Haila 2005) or, from a post-colonial perspective, considered as part of the Global North and thus compared with the cities of the Global South (Gugler 2004, Grant/Nijman 2006, Robinson 2006, Simon 2006, Haferburg/Oßenbrügge 2009, Diez/Scholvin 2017).

Despite ongoing uncertainty as to whether the "characteristic features" of the European city, such as local government, the influence of active residents and the erosion of the dichotomy between the public and the private, are still relevant to current developments (Sennett 1983, Siebel 2015), by the same token, it can be argued

that the ability to respond flexibly to ongoing processes is also a key feature of the European city (Sennett 2018, BBSR 2010).

Our analysis of the impacts of CAM on the European city centres on how urban society, urban planning and urban policy are responding to changing economic, ecological, social and architectural objectives. This subsequently means that the European city must be understood as multidimensional and considered from an interdisciplinary perspective.

One significant aspect regarding policy and planning in the European city is the relatively high level of autonomy enjoyed by city policymakers and planners as part of the subsidiarity principle. This sees the responsibilities of national government devolved to the regional/local level and financially covered through the provision of transfer payments by the state (Siebel 2004). However, binding cities to states in this manner also makes them susceptible to welfare state restructuring during which responsibilities are transferred to those "lower down" without the relevant funding being secured (Jessop 1992, Brenner 2004). Against this backdrop, the European city also became the "entrepreneurial city" (Harvey 1989, Häußermann 2001). Klaus von Dohnanyi, a former mayor of Hamburg, was the first mayor of a German city to refer to his jurisdiction as an *Unternehmen* or "enterprise" (Dohnanyi 1983, Dangschat 1992).

The current relevance and significance of financial autonomy can also be seen with regard to the potential fiscal impacts of CAT, which were examined as part of the Vienna project (Soteropoulos et al. 2018b; see also Chap. 4.3). Parallel to the outlined urban development trends, we can examine European city transport and mobility planning since the Second World War by dividing developments into three different stages (see Chaps. 3.2.1 to 3.2.3 as well as Figs. 3.2.4 and 3.2.5).

3.2.1 DESTRUCTION AND REBUILDING – STAGE 1

Large-scale destruction during World War II paved the way for a departure from the industrial, workers' cities of the past that were characterized by high-density housing, hardship and a struggle for survival. Light, air and sunshine would be the order of the day, resulting in the concept – based on the Athens Charter, a largely for-

Figure 3.2.1: Dimensions of the European city

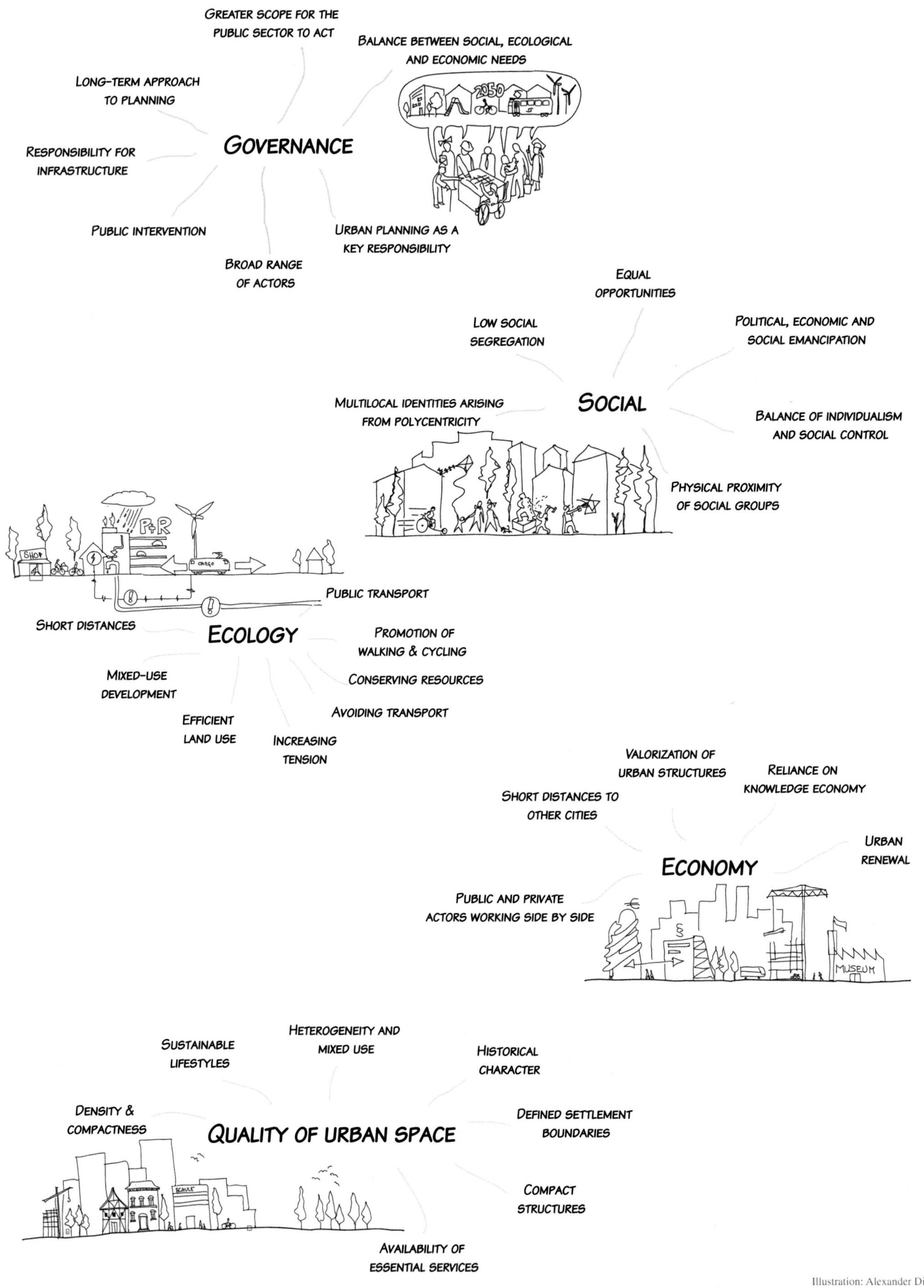
GREATER SCOPE FOR THE PUBLIC SECTOR TO ACT
BALANCE BETWEEN SOCIAL, ECOLOGICAL AND ECONOMIC NEEDS
LONG-TERM APPROACH TO PLANNING
GOVERNANCE
2050
RESPONSIBILITY FOR INFRASTRUCTURE
PUBLIC INTERVENTION
BROAD RANGE OF ACTORS
URBAN PLANNING AS A KEY RESPONSIBILITY
EQUAL OPPORTUNITIES
LOW SOCIAL SEGREGATION
POLITICAL, ECONOMIC AND SOCIAL EMANCIPATION
MULTILOCAL IDENTITIES ARISING FROM POLYCENTRICITY
SOCIAL
BALANCE OF INDIVIDUALISM AND SOCIAL CONTROL
PHYSICAL PROXIMITY OF SOCIAL GROUPS
SHOP
P+R
CARGO
PUBLIC TRANSPORT
SHORT DISTANCES
ECOLOGY
PROMOTION OF WALKING & CYCLING
MIXED-USE DEVELOPMENT
CONSERVING RESOURCES
AVOIDING TRANSPORT
EFFICIENT LAND USE
INCREASING TENSION
VALORIZATION OF URBAN STRUCTURES
RELIANCE ON KNOWLEDGE ECONOMY
SHORT DISTANCES TO OTHER CITIES
URBAN RENEWAL
ECONOMY
PUBLIC AND PRIVATE ACTORS WORKING SIDE BY SIDE
MUSEUM
SUSTAINABLE LIFESTYLES
HETEROGENEITY AND MIXED USE
HISTORICAL CHARACTER
DENSITY & COMPACTNESS
QUALITY OF URBAN SPACE
DEFINED SETTLEMENT BOUNDARIES
SCHULE
COMPACT STRUCTURES
AVAILABILITY OF ESSENTIAL SERVICES

Illustration: Alexander Diem

gotten pamphlet created during a meeting of members of the CIAM (Congrès Internationaux d'Architecture Moderne) under the leadership of Swiss architect Le Corbusier – of *Die aufgelockerte und gegliederte Stadt* (the articulated and relaxed city, Göderitz et al. 1957; see Fig. 3.2.2). In addition to the damage caused by the conflict, the new focus on functional separation that emerged from the proverbial and literal ashes contributed to the further "destruction" of the traditional European city, i.e. its urban planning structures, ideas of urbanity and social cohesion.

The post-war period also saw rapid growth in car-based mobility. First mass-produced by Henry Ford, who created the "universal car" or Model T, which was to be affordable for factory workers, the car was later used to political ends by the Third Reich in the form of "Volkswagen" (the people's car). Now, in peacetime, the private vehicle was being talked up as a symbol of the economic miracle and an embodiment of Europe's road to recovery. The conditions were such that governments were willing to create vast amounts of space for cars and to develop cities with automobiles in mind based on the *autogerechte Stadt* (car-friendly city) proposed by Hans Bernhard Reichow, who also developed a corresponding conceptual urban design plan in 1959, albeit one which also took different modes of transport into consideration. In the years that followed, the notion of a car-friendly city grew to extremes. Cars needed to be able to travel largely unimpeded, and so any hindrances, such as buildings, pedestrian crossings or even tram lines, were removed. Corridors were created for urban

motorways together with extensive traffic junctions, with urban cityscapes having to make way.

This urban planning model for the reconstruction and expansion of settlement structures came to define western European cities, but also cities in socialist-led countries, for roughly five decades (Goldzamt 1973). Large housing estates were created next to office blocks, shopping centres were built alongside universities and other institutes of education, all separated by a green belt and connected via car-friendly roads. The use of cars, which rose considerably from the end of the 1960s, also provided the basis for suburbanization: the movement of young, upwardly mobile families and household-based services, and then, later, offices and light industry, to peri-urban areas (Friedrichs 1978, Brake et al. 2001).

In the late 1960s and early 1970s, views started to change. In 1965, German psychoanalyst Alexander Mitscherlich published a book titled *Unwirtlichkeit der Städte* (The Inhospitality of Our Cities) in which he criticized Germany's urban planning and renewal strategies, which were centered purely on functionalistic principles. The Deutsche Städtetag (German Association of Towns and Cities) headed by the then Mayor of Munich, Hans-Jochen Vogel, eventually called for German cities to be saved immediately (DStT 1971), and the Club of Rome published a report on *The Limits to Growth* in 1972. However, the oil crisis finally made clear that a rethink of how resources were used was desperately needed, all of which led to a fresh approach and return to the values and characteristics of the European city.

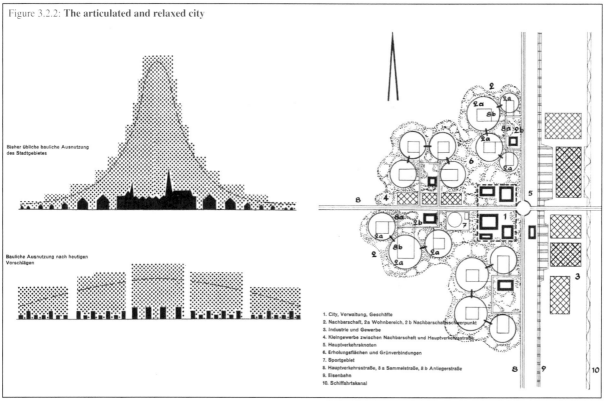

Figure 3.2.2: **The articulated and relaxed city**

Bisher übliche bauliche Ausnutzung des Stadtgebietes

Bauliche Ausnutzung nach heutigen Vorschlägen

1. City, Verwaltung, Geschäfte
2. Nachbarschaft, 2a Wohnbereich, 2 b Nachbarschaftsschwerpunkt
3. Industrie und Gewerbe
4. Kleingewerbe zwischen Nachbarschaft und Hauptverkehrsstraße
5. Hauptverkehrsknoten
6. Erholungsflächen und Grünverbindungen
7. Sportgebiet
8. Hauptverkehrsstraße, 8 a Sammelstraße, 8 b Anliegerstraße
9. Eisenbahn
10. Schiffahrtskanal

Figure 3.2.3: **Friedrich-Engels-Platz in Leipzig following its reconstruction in 1971**

A car-friendly city created by segregating the modes of transport: separated, low-conflict spaces for pedestrians, cars and trams, as well as a spacious design of the entire traffic area

Source: German Federal Archives

3.2.2   CAUTIOUS URBAN RENEWAL – STAGE 2

The second stage of post-war urban development saw the start – partly triggered by huge protests – of policymakers replacing their previous strategy of urban redevelopment with one centred on "cautious urban renewal". Although this approach initially consisted solely of structural decisions, i.e. whether a building was "worthy of preservation", the passing of the Urban Development Promotion Act in 1971 made it mandatory for residents to be involved in redevelopment projects. At the end of the 1970s, during the redevelopment of Berlin's Kreuzberg district, "12 principles of urban renewal" were developed (Hämer 1990) that were adopted by district representatives before going on to become a model for the whole of Berlin in 1984 (thanks to International Building Exhibitions (IBA)) and, ultimately, urban regeneration across the entire nation.

Pursued alongside state regulations (such as tenant protections, building refurbishment and housing subsidies, and tax deductions for privately owned housing) and the architectural transformation of public space by bringing in traffic calming measures, this "cautious urban renewal" approach encouraged an increased demand for inner-city living, which in large cities, such as Munich, Hamburg and Düsseldorf, led to a process of gentrification that has been ongoing since the late 1970s (Dangschat 1988). Transport development was now boosted by the rapid expansion of public transport (esp. rail); however, this took place without first reducing the space afforded to cars.

3.2.3   THE LIVEABLE CITY – STAGE 3

The second stage, which tried to incorporate more balance into urban planning, made way for a third stage that was characterized by renewal, improvement of public space, greater quality of life by reducing emissions (not just greenhouse gases but also noise) and encouragement of – and demand for from certain social groups – active mobility by promoting walking, cycling and travel by scooter (Jones 2017). Schemes to improve the local living environment and calm traffic were introduced at the federal and state level that were later combined to form a broader concept as part of the "Socially Integrative City" scheme adopted by German federal and state governments. This scheme placed integrated planning approaches that aimed to promote vibrant neighbourhoods and social cohesion centre stage.[1] Increasing ecological and (urban) climate issues and challenges, as well as a growing interest, among certain parts of the urban population, in eating sustainably and ethically, healthy living, well-being and a high quality of life (LOHAS = Lifestyles of Health and Sustainability) added increased momentum and zeal to achieve sustainable urban development targets that continue to this day.

Policymakers and planners responded to this shift by redesigning and revolutionizing public space, introducing traffic calming measures and expanding cycling lanes as well as removing parking spaces. House builders had to adhere to a growing number of restrictions that stipulated the construction of low-energy, accessible homes, and mobility concepts were also developed that revolved around residents eschewing private cars and thus reducing mobility costs. At the same time, efforts were once again

Figure 3.2.4: **Stages of transport and mobility planning and policy in the European city**

	STAGE 1 Accommodating traffic growth	STAGE 2 Encouraging modal shift	STAGE 3 Promoting liveable cities
CHARACTERISTICS ACCORDING TO JONES	• Rapid growth in car ownership (among the wealthy) • Focus on the vehicle and infrastructure • Economic growth becomes the objective • Lack of investment in walking and cycling	• Negative social and environmental effects become apparent • Regulatory approaches and public influence • Improvement of public transport • Parking management, restricted access	• Focus on liveable spaces and sustainable forms of mobility • Aim to increase the quality and liveability of urban areas • Stricter controls and socio-science approaches • Reclaiming public space • Car ownership starts to decline
PLANNING PARADIGMS	• The articulated and relaxed city (Göderitz 1957) • Car-friendly city (Reichow 1959) • Traffic in Towns (Buchanan 1963) • Athens Charter (CIAM 1933)	• 12 principles of cautious urban renewal (Hämer 1990) • IBA Berlin (1984) • Traffic-calming measures in residential areas: large-scale trials (DE), woonerf concept (NL)	• Association of German cities and towns, 2018 • SUMP (Sustainable Urban Mobility Plans) • Guidelines for Urban Road Design (RASt) • Leipzig Charter, 2007

Source: Characteristics according to Jones (2017)

stepped up to encourage social diversity when housing initial occupants, and neighbourhood management systems were introduced in so-called "problem" areas.

In terms of transport and mobility planning, increased focus was given to the shift from motorized private transport (MPT) to public transport, cycling and walking (ecomobility). The expansion of multimodal traffic concepts became the focus of urban and mobility development. Moreover, recent years have seen the advent of connectivity between various modes of transport thanks to apps and digital platforms that offer a wide range of transport options, a comprehensive ticketing system, calculation of costs and additional information (MaaS). The current objective of transport and mobility policy is to help achieve a largely car-free multimodality in cities. However, at the same time, we can observe a rise in delivery vehicles in cities, which can be attributed to the increasingly influential role played by digital commerce.

3.2.4 THE INFLUENCE OF CONNECTED AND AUTOMATED VEHICLES ON TRANSPORT AND MOBILITY POLICY

We now turn to the question of how the introduction of CAT will influence transport and mobility policy in the European city. CAT is generally perceived as a positive development (STRIA 2019), but such evaluations rarely discuss which (urban) development measures, regulations and monitoring systems will be required for its successful implementation. This in turn raises the question of whether CAT is in line with and helps achieve the objectives of current developments in urban and mobility planning as set out in Stage 3, and whether, given CAT's need for separate and ample space, and individual and often protected lanes, it will make new traffic structures

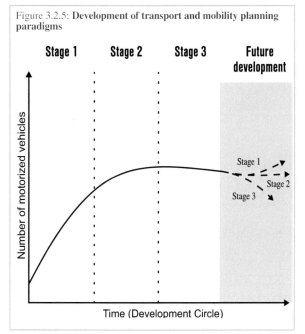

Figure 3.2.5: **Development of transport and mobility planning paradigms**

Source: AVENUE21 based on Jones (2017)

necessary (Rupprecht et al. 2018). Lastly, it is assumed that CAVs will result in an increase in traffic volume and therefore feared that the European city could be redesigned to accommodate CAVs (return to Stage 1; Jones 2017, Dangschat 2018, Rupprecht et al. 2018). Figure 3.2.4 illustrates that future governance will heavily influence how transport and mobility planning is adapted in the years to come, as will the objectives that are chosen within this context (see scenarios in Chap. 5).

---

1   At the European level, the Leipzig Charter, which was instigated by Germany, saw the creation of a Europe-wide model for sustainable development; the charter has become binding for European urban development (BMVBS 2007). The objectives outlined in this charter contrast markedly with the guidelines issued in the Athens Charter.

3.3
NEW MOBILITY:
DEVELOPMENTS, OPPORTUNITIES AND RISKS[1]

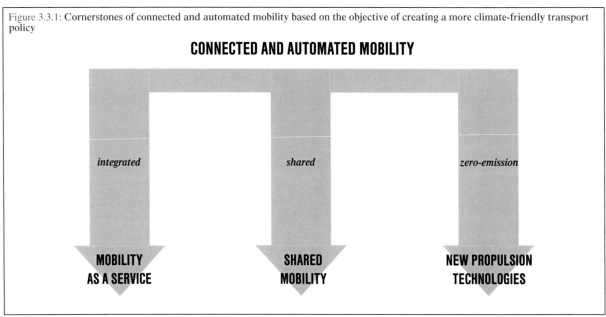

Figure 3.3.1: Cornerstones of connected and automated mobility based on the objective of creating a more climate-friendly transport policy

Source: AVENUE21

Megatrends are an effective way to narrow down the potential future pathways urban planners and policy-makers may take[2] and to provide useful indicators for cutting-edge research (WBGU 2011). The current megatrends – and their impact on (automated) mobility – are discussed in more detail in Chapter 3.1. Amidst the current push to create a more environmentally friendly transport policy (Chap. 1), simply automating vehicles does not go far enough given the increasingly urgent need for action as we face up to the reality of climate change. What will be crucial is whether automated vehicles can be developed that are low emission (or emission-free) and whether they will be embedded within an integrated Mobility as a Service concept as shared mobility (Lennert/Schönduwe 2017). We will thus examine the MaaS concept, as well as key elements such as shared mobility and new propulsion technologies, against the backdrop of automated driving, and discuss the future development pathways these technologies may take.

3.3.1 MOBILITY AS A SERVICE

Mobility as a Service (MaaS) is a concept whereby public and private transport services (as well as different forms of mobility, including automated vehicles) are combined with a single digital access portal (a platform or an app) in order to offer mobility solutions that are tai-

lored to meet people's individual needs (EPOMM 2017; Jittrapirom et al. 2017: 14). Connected and automated vehicles aid the development of MaaS by continually blurring the lines between classic public transport and MPT thanks to the automation of vehicles, resulting in increasingly flexible and independent movement (Lenz/ Fraedrich 2015: 189; Bruns et al. 2018: 12). The technological development of CAVs opens new possibilities for the development of business models, which could enable new suppliers to gain entry to the market. Due to ongoing automation and connectivity, it is possible that disruptive developments will take place in the mobility sector and that existing services will undergo further transformation in the years to come (Gertz/Dörnemann 2016: 5). Key components of a MaaS solution are (Jittrapirom et al. 2017: 16; Lund 2017):

- *MaaS operators/integrators*
  These actors sell a comprehensive service to the end consumer, handle activities such as customer management and carry out marketing strategies. MaaS services can be either private or public, or a mix of both. These mixed models are called PPP (public–private partnership) or PPPP (public–private–people partnership). The latter can typically also be expanded to include peer-to-peer sharing and social dimensions (Aapaoja et al. 2017: 9–11). The challenges of using MaaS in

practice often arise during the search for a suitable operator structure: on the one hand, businesses solely offering a platform have no control over, and no responsibility for, the individual services. On the other, public transport operators and private mobility providers have an interest in promoting their own mobility services (Smith et al. 2017: 8).

- *Cooperation between various mobility providers*
  Cooperation between various mobility providers (such as car and bike-sharing as well as public transport and taxi companies) in a horizontal integration model is vital for a successful system (Joschunat et al. 2016: 70; Li/Voege 2017). To enable door-to-door mobility, the aim should be to create a pool of supply-based services (public transport operating on an interval timetable) and demand-based services (e.g. bike sharing, connected and automated ride and car sharing; Lund 2017). Most notably with regard to the first and last mile, automated vehicles are seen to offer huge potential as a shuttle system to carry travellers to public transport hubs in urban and rural areas (BMVIT 2016c, Ohnemus/Perl 2016). As there is now a bigger choice of transport options and mobility services, MaaS operators are better placed to meet customers' various needs and preferences (Goulding/Karmagianni 2018: 2).

- *Mobility platforms (information and communication technology)*
  The core components, such as information on alternative forms of mobility and bookings as well as payment and billing for used mobility services, are all managed on one platform

(vertical integration; Joschunat et al. 2016: 70). According to Sochor et al. (2017: 193–196), the scope of vertical integration can be divided into different levels: Level 0 (no integration), Level 1 (integration of information), Level 2 (integration of booking and payment), Level 3 (integration of the service offer) and Level 4 (integration of societal goals, incentives).

At present, MaaS approaches are being implemented internationally in different contexts[3]: in 2014, a mobility app was tested in 70 households in Gothenburg as part of the "Go:Smart/UbiGo" pilot; the "UbiGo app"[4] was also trialled in Stockholm in 2018. An important finding that resulted from the first pilot project in Gothenburg was that MaaS can lead to a change in mobility behaviour and to higher user satisfaction (https://ubigo.me/). The "Whim app" (https://whimapp.com/) was introduced in Helsinki in 2016, Austria has become familiar with a functionally and modally integrated MaaS system thanks to the countrywide "SMILE – einfach mobil" project (2012–2015), and residents of Vienna have had access to the "WienMobil" app since 2017.

A growing percentage of the population is now taking advantage of multimodal mobility, i.e. individuals are using different forms of transport to get from A to B (Busch-Geertsema et al. 2016: 757). Smartphones and apps are being used to find out the best way (e.g. the fastest or most convenient route) to complete a journey – on foot, by bike, public transport, car sharing, etc. – and this is particularly true in urban areas and among younger members of the population (BMVIT 2019). The (technological) development of platforms (individualization of services, customization options, e.g. to suit an individual's personal routine) as well as the integration of

Figure 3.3.2: **MaaS – opportunities, risks and obstacles**

OPPORTUNI-TIES	• Creation of competitive, sustainable alternatives to private cars and a reduction of MPT use (see Lund 2017, Holmberg et al. 2016) • Improved efficiency of existing mobility services and public transport, including in less densely populated areas (Gertz/Dörnemann 2016, Hoadley 2017, Bösch et al. 2018) • Development of an inclusive mobility system as MaaS can be adapted to personal needs (personalization of service; Hoadley 2017)
RISKS	• Exclusion of the less tech-savvy by digitalizing transport services ("digital gap"; Hoadley 2017) as well as those who cannot afford to access such services (mobility poverty) by introducing business models, i.e. private operator structures (Pangbourne et al. 2019) • Socio-spatial inequalities that arise when commercially designed operator structures lead to MaaS being offered exclusively in densely populated urban areas and not in less densely populated locations (Alberts et al. 2016) • Rebound effects, e.g. when an imbalance of transport modes occurs (Eckhardt et al. 2018), resulting in those previously unable to use motorized vehicles being granted access (Datson 2016, Durand et al. 2018)
OBSTACLES	• Casual use of the term "MaaS" – the objective of MaaS must be to achieve MaaS Levels 3 and 4 (Harms et al. 2018) • Even greater uncertainty about the effects of MaaS at the individual (mobility behaviour, everyday integration) and social (e.g. social and ecological sustainability; Durand et al. 2018) levels • MaaS will pose significant challenges for governance structures, e.g. lack of provisions for MaaS in public strategies, the availability of data (Big Data) resulting in power being transferred to private actors, the risk of innovations being outsourced to the private sector (Pangbourne et al. 2019)

additional services make it possible to reach new target groups. What is noticeable is that previous socio-ecological ideals are increasingly being superseded by customers' pragmatic attitudes to mobility: flexible service provisions appeal to highly mobile individuals with a multitude of choices who want to ensure they have several travel options at their disposal (Maertins 2006).

The level of demand for MaaS depends on a number of factors (Harms et al. 2018: 23–24):

- *Mobility behaviour*
  Studies show that car owners[5] who use their vehicles very often (on four or more days per week) and never or rarely use public transport are the least inclined to use MaaS (Ho et al. 2017). Experience with (intermodal) public transport, however, increases the likelihood of an individual using MaaS as well as other forms of transport.

- *Route characteristics*
  MaaS holds unique potential when it comes to leisure travel and journeys to irregular, unknown destinations: integrating additional information and making it accessible to the user (Harms et al. 2018: 23).

- *Digital skills*
  Young, tech-savvy adults tend to use MaaS more compared to older generations or those less familiar with technology (Kamargianni et al. 2018).

- *Socio-demographics*
  Households with two or more small children show less interest in MaaS than other households (Haahtela/Viitamo 2017; Ho et al. 2017). Given the fact that social structures crucially influence mobility and how it can be accessed, it is important that any future MaaS research not only analyses individuals but also considers familial structures (Haahtela/Viitamo 2017).

- *Aspects of (mobility) culture*
  How much a society is "service-orientated" plays a particularly important role (Haahtela/Viitamo 2017).

These points show that within a society, several different factors are at play, which determine the level of accessibility of MaaS services. Against this backdrop, MaaS services need to be introduced in a way that targets specific groups and avoids replicating existing social inequalities (e.g. the "digital divide"; Durand et al. 2018).

3.3.2 SHARED MOBILITY

Shared mobility is just one part of the sharing economy and concerns the shared use of mobility services (BMVIT 2016c: 12). Shared mobility is a part ownership-based, part public mobility model and grants users access to various means of transport without the need for ownership (Kollosche/Schwedes 2016: 26).

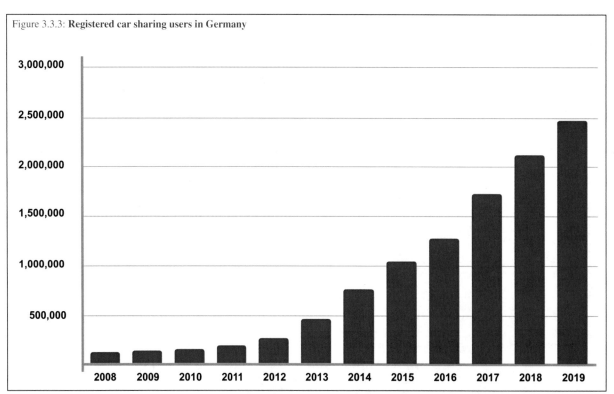

Figure 3.3.3: **Registered car sharing users in Germany**

Source: AVENUE21 based on Bundesverband CarSharing (2019), accessed via Statista

As shared mobility increases the number of passengers on a journey and in a specific vehicle, it is considered vital in helping to realize the climate-friendly and re-source-efficient mobility model of the future. Within the context of automated vehicles too, shared use will be a vital prerequisite to significantly reducing the number of private cars, enabling more efficient use of the existing infrastructure and allowing an improved quality of life through the recovery and repurposing of space dedicated to roads (BMVIT 2016c: 60; see Chap. 4.3).

Shared mobility services are usually booked and in-voiced via an app and/or an internet platform. In recent years, the mobility sector has seen the addition of a large number of sharing services with different organi-zational structures and motives (Scholl et al. 2013; BM-VIT 2016c), e.g. commercial (Business-to-Consumer – B2C, Business-to-Business – B2B), non-commercial (Consumer-to-Consumer – C2C) and public (Govern-ment-to-Consumer – G2C). Currently, shared mobility services are primarily offered via two different systems: they are either station-based or free floating. The grow-ing appeal of shared mobility can mainly be attributed to free-floating systems, which, thanks to CAVs that can travel the few metres back to the pick-up location, could be set to become more significant both in absolute and in relative terms (Shaheen/Chan 2016: 577).

However, shared mobility continues to be used in a so-cially selective manner: customers are more likely to be male and tend to be younger on average than the rest of the population; users also have a comparatively higher level of education and a higher income (Böhler et al. 2007, Kopp et al. 2015, Riegler et al. 2016, Hülsmann et al. 2018).

Sociocultural factors favouring shared mobility are:

■ A shift in values characterized by the fading symbolic value of property (Botsman 2013, Owyang et al. 2014, Priddat 2015).

■ Sharing is associated with modern values, a higher level of freedom as well as greater flexi-bility and independence (Harms 2003).

■ Growing general awareness of the ecological consequences of individual actions, although the "green image" (Steding et al. 2004, Gossen 2012, Lindloff et al. 2014) is now becoming less important.

■ Everyday compatibility and pragmatic arguments are gaining ground (Loose 2010, Lindloff et al. 2014; e.g. convenience, flexibility, loan points can be easily reached, service is simple and straightforward to use, costs less).

Mainly two forms of shared mobility are relevant within the context of CAT: car sharing and ride sharing. In both examples, connected and automated vehicles offer the potential to create systems that are more affordable and efficient (Bösch et al. 2018: 82). The following looks at both services in detail.

Figure 3.3.4: **Shared mobility – opportunities, risks and obstacles**

OPPORTUNI-TIES	• Data analysis of how the rising level of automation has affected mobility behaviour and, subsequently, the scope to optimize services (Freese/Schönberg 2014)
	• Savings and more efficient use of resources achieved through connected and automated vehicles (Bösch et al. 2018)
	• Expansion of shared mobility services by diversifying the types of vehicles available (e.g. e-cars; BMVIT 2016c)
	• Using instead of owning: a shift in attitudes tends to lead to further growth of shared and connected forms of mobility (BMVIT 2016c) driven by information and communication technologies, digitalization and cultural processes of change (Alberts et al. 2016)
RISKS	• Lack of oversight of the sharing services on offer, lack of connectivity and integration (if the service is not embedded within a MaaS system), substantial effort required from users (BMVIT 2016c)
	• Relocation challenges in free-floating systems: vehicles are stuck in "cold spots" that are unattractive to users, idle periods are not profitable for operators (Weikl/Bogenberger 2013)
	• New forms of shared mobility (e.g. e-scooters) trigger conflicts regarding use in public space (Riegler 2018)
OBSTACLES	• Everyday mobility shaped by routine (Scheiner 2009)
	• Anxieties and concerns about "others" keep demand low for the shared use of driverless small vehicles (Salonen/Haavisto 2019)
	• Accessibility (distance to the vehicle) and availability of vehicles – connecting and automating the systems could counteract this (BMVIT 2016c)
	• Organizational aspects and corporate strategies impede the development of an integrated information and communication platform (MaaS; BMVIT 2016c)
	• Partial lack of predictability and security resulting from the high degree of flexibility offered by shared mobility (Vogel et al. 2014)

Commercial car sharing (B2C) is already well established in Europe's major cities. In recent years, free-floating systems in particular have reached a growing number of major cities where the respective networks have gradually been expanded and the modes of transport on offer have been diversified through the addition of bike sharing, e-moped sharing and e-scooter sharing. Although (car) sharing is seeing substantial growth,[6] the number of car-sharing users as a share of the total population remains rather low (see Fig. 3.3.3).

Austria is one of the countries where user figures remain low but the trend is heading upwards: of Austria's approx. 3.9 million private households, only around 100,000 use car sharing (BMVIT 2016a in VCÖ 2018b). A study published by PwC in 2018 predicts that by 2030 more than one in three kilometres driven in Europe will be completed using some form of shared mobility. The number of private cars that could be replaced by car sharing depends on the respective system, the overall conditions within the transport system in the various cities and the aspects concerning (mobility) culture. Scenarios and prognoses thus calculate vastly different figures with regard to the potential reduction in greenhouse emissions. For example, in a study for the city of Munich, the authors estimate that one free-floating car sharing vehicle could replace 3.6 private cars (Schreier et al. 2015); in a study on another German city, Bremen, they predict that each station-based car-sharing vehicle could replace 16 private cars (Schreier et al. 2018). Generally speaking, the large-scale operators believe the future of car sharing lies in automation and the use of electric mobility. For instance, there are plans for the entire Car2Go fleet to be fully autonomous and electric by 2030 (Stüber 2018).

In addition to sharing the actual mode of transportation ("good sharing"), there are other forms of sharing where a vehicle is used simultaneously by different individuals: depending on the providers, the services on offer range from ride pooling (carriers, rental and taxi companies that are obligated to ensure continuous service), ride sharing (e.g. Blabla Car) and ride selling or ride hailing (commercial platform operators, e.g. Uber; cf. Sommer 2016). At present, there are a number of barriers particularly with regard to (C2C) ride sharing from the user's perspective (accessibility, safety, proximity to strangers; Nielsen et al. 2015). To that effect, connected and automated ride sharing could present a huge opportunity to increase usage by enabling a higher level of flexibility (Bruns et al. 2018: 22).

3.3.3 NEW PROPULSION TECHNOLOGIES

As resources are set to become even more scarce and we face the evident threat of climate change, it is vital to also discuss CAVs within the context of alternative fuels and new propulsion technologies.[7] A growing number of alternative solutions are being developed to compete with fossil fuels (Kollosche/Schwedes 2016: 19–20). These are:

- battery-powered electricity (generated by renewables),

- electric motors powered directly (via cables or induction),

- different generations of biofuels,

- fuel cell cars powered by hydrogen (see UBA 2015 for more on the advantages and disadvantages of different vehicle generations).

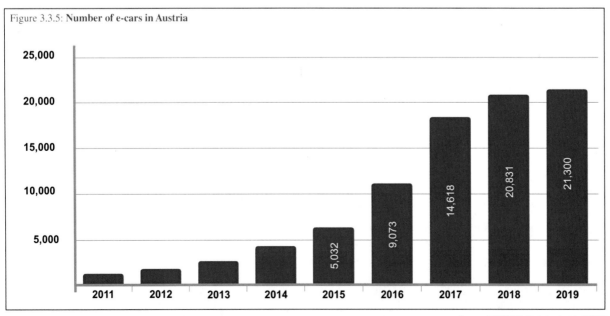

Figure 3.3.5: **Number of e-cars in Austria**

Source: AVENUE21, Statistics Austria (2019b)

To be able to use these new types of fuel, new propulsion systems need to be designed and produced. Which adapted propulsion systems and respective fuels will prevail on the market depends on their level of efficiency (e.g. cost effectiveness, environmental compatibility and practical applicability; Kollosche/Schwedes 2016: 19–20):

- *Battery Electric Vehicles (BEV)*: electric motor with a battery, can be recharged from the grid; limited distances of 200–400 km.

- *Fuel cell vehicles* generate the necessary power with the aid of a fuel cell, which powers an electric motor; hydrogen is used as fuel; no grid connection required; average distances of 400–600 km.

- *Fuel cell hybrid electric vehicles*: hydrogen is converted into electrical energy in the fuel cells; vehicles thus feature both fuel cells and a battery.

- *Hybrid vehicles*: a combination of a classic combustion engine and an electric motor; no need to connect to the grid (with the exception of plug-in hybrid electric vehicles – PHEV).

Politicians at both national and international levels consider electromobility to be key to decarbonizing transport (BMVIT n.d.; European Commission 2018). Although the number of registered electric vehicles has been rising gradually, overall the transition to electric models has been slow (AustriaTech 2018). If alternative fuels and propulsion technologies are to achieve substantial market penetration, they will first need to find a certain level of (growing) acceptance among end consumers. But as the registration figures show, there is quite a high degree of scepticism. This is because, on the one hand, the public are not fully aware of the advantages of these technologies, while the focus is often placed on the constraints that currently exist (high purchasing costs, limited range, poor network of charging stations; Bobeth/Matthies 2016). However, these arguments are also often used to justify emotionally driven reservations. And

even though the range of such vehicles would suffice for the majority of journeys (Kollosche/Schwedes 2016: 19–20), these are the arguments made time and time again. On the other hand, users sometimes consciously choose to ignore the fact that their (fossil fuel-based) mobility and the resulting emissions are contributing to climate change.

However, in sharing systems, electric vehicles are met with a significantly higher level of acceptance and appear to have greater appeal: they are seen to be more environmentally friendly and almost just as practical as conventional vehicles (Hülsmann et al. 2018: 120). Car sharing thus plays a vital role in breaking down psychological barriers to e-mobility and in allowing low-threshold contact points (Hülsmann et al. 2018: 120). There are now e-car sharing providers in almost every Austrian federal state (e:mobil 2018), although community-based (stationary) e-car sharing has mostly been established in rural areas (partly through targeted "klimaaktiv mobil" funding; klimaaktiv 2017).

One particular political target – that of increasing the number of e-vehicles – requires a significantly more radical expansion of the charging infrastructure in residential buildings, as well as of renewable energies and network capacity. In addition to boosting the availability of charging stations, it will also be important to improve charging times in order to increase public acceptance of e-mobility. It is thus crucial that the time spent waiting for a vehicle to charge can be made more enjoyable; here there is still a significant lack of suitable solutions (Ebert et al. 2012).

With this in mind, the task at hand is also to increasingly utilize the synergies between electromobility and automated driving. The process of automation will thus be able to break down some of the obstacles that prevent individuals from using electromobility (e.g. users' fear of not being able to travel far, access to charging infrastructure, management of charging time). Connected and automated vehicles are able to manage these aspects au-

Figure 3.3.6: **New propulsion technologies – opportunities, risks and obstacles**

OPPORTUNITIES	• Low-threshold access to new propulsion technologies enabled by shared mobility (VCÖ 2018b; Hülsmann et al. 2018)
	• Increased quality of life in cities thanks to fewer emissions (pollutants and noise pollution; VCÖ 2011)
RISKS	• Path dependency resulting from the current focus on e-mobility (Fischedick/Grunwald 2017)
	• User scepticism towards new propulsion technologies (Kollosche/Schwedes 2016)
	• Concerns about road safety (e.g. alternative propulsion systems producing low noise levels; Ingenieur.de 2018)
OBSTACLES	• Industrial strategies put forth by policymakers that restrict the further development of technologies to help produce alternative fuels and propulsion systems (Kollosche/Schwedes 2016)
	• E-mobility and its expansion are tied to infrastructure development (charging stations) and the standardization of charging plugs and access schemes, communication protocols and solutions for billing systems (e.g. Ebert et al. 2012)

tonomously based on real-time journey demand (Chen et al. 2016). In order to take advantage of the technological synergies available, RWTH Aachen is currently running a project (UNICARagil, www.unicaragil.de) to develop a modular and scalable vehicle concept for electric-powered automated vehicles that can be adjusted flexibly to suit a wide range of applications in logistics and passenger transportation.

In conclusion, it is clear that any further examination of the potential CAVs hold to help bring about a shift in mobility must also consider the three trends mentioned here – MaaS, shared mobility and new propulsion technologies. For this reason, we actively included the three factors when developing our scenarios (see Chap. 5).

———

1   Vanessa Sodl (a researcher in Transportation System Planning, TU Wien) played a considerable role in helping compile this report and her knowledge, in particular on MaaS and new propulsion technologies, was invaluable to our work.

2   However, megatrends represent, at best, general frameworks that take effect on a global scale and are thus in no way deterministic development trajectories.

3   Even if the concept of MaaS integration is only comprehensively embedded in a handful of projects.

4   The app combines public transport, car sharing, car hire and taxis in one intermodal mobility service. Each household selects a flexible monthly subscription and the account is shared by all members of the household.

5   In 2017, there were 371 private cars per 1,000 inhabitants in Vienna (the figure stood at around 500 private cars per 1,000 inhabitants in other Austrian federal state capitals due to less effective public transport systems). The figure was 349 in Munich and 346 in Hamburg, with a slightly higher number for Berlin (384). In Europe's developed cities, 300 private cars per 1,000 inhabitants has become somewhat of a benchmark that symbolizes a considerably lower private car stock (ORF 2018).

6   In recent years, however, the trend seems to have reversed somewhat, which can also be seen in the merging of different providers.

7   Yet it is assumed that any changes are likely to be in small, gradual steps, which means that the combustion engine will still be the dominant form of propulsion until the year 2040 (Bukold 2015: 3). On the other hand, political decisions, such as those made in Norway, France and also China, indicate that this shift could take place at a faster pace in at least some regions.

3.4

EXPERTS' IMPACT ASSESSMENT OF CONNECTED AND AUTOMATED MOBILITY[1]

3.4.1 SURVEY AIM AND METHODOLOGY

In the autumn of 2017 and the winter of 2018/2019, experts from a broad range of fields that included urban development, mobility planning and technology development were invited to participate in two online surveys as part of the AVENUE21 project. The aim of both surveys was to ascertain the existing level of knowledge within a range of academic, planning and business settings concerning the link between connected and automated road vehicles and the development of European urban regions – an area which, according to Fraedrich et al. (2018), is still under-researched. Within the project as a whole, these two surveys would also support the concurrent creation of scenarios (Chap. 5).

Data were collected using standardized electronic surveys; a total of approx. 980 individuals were approached and invited to take part. When selecting a list of prospective experts, our team made sure to choose specialists with a broad range of expertise. We contacted individuals who are conducting research in different academic fields (technology, social sciences, business or legal sciences) or working for mobility service pro-

viders as well as experts involved in development and public administration, land-use planning, consultancy or politics. When conducting both surveys, we contacted specialists both in German-speaking countries (i.e. using a German questionnaire) and in the rest of Europe (predominantly from the Netherlands and the United Kingdom, who received an English questionnaire). The majority of respondents for both surveys were from German-speaking countries.[2]

We received over 200 responses to the surveys, i.e. our response rate was above 20% (first survey: 211, second survey: 216). Although this may be a relatively high number for a survey of this kind, given the non-response rate of just below 80%, we should expect some of the results to be distorted.

As this project examines the issue of CAT and the impact it has on the European city (in terms of impacts on governance, architecture and urban development as well as urban society), the urban and transport planning professions are comparatively overrepresented in these surveys, while those specialists who work more specifically with technological systems tend to be underrepresented.

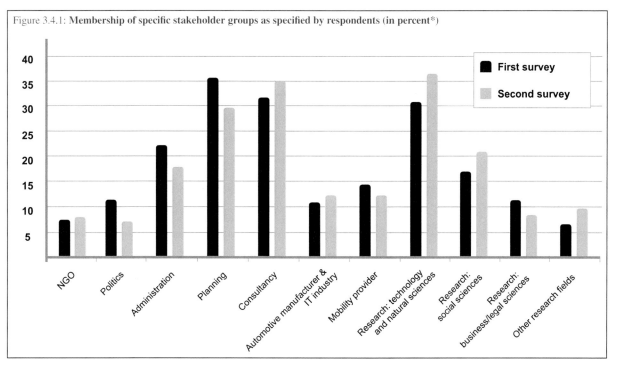

Figure 3.4.1: **Membership of specific stakeholder groups as specified by respondents (in percent*)**

* As respondents were able to select multiple answers, the percentage values add up to more than 100%
Source: AVENUE21

The respondents' answers regarding their professional field (see Fig. 3.4.1) indicate that a particularly high number of professionals from the fields of "planning", "consultancy" and "technology or natural sciences research" (each accounting for between 30% and 37% of participants) took part. In contrast to the first survey, professionals active in the field of technology or natural sciences research interestingly formed the largest group of participants in the second survey.

3.4.2   FIRST SURVEY: STAKEHOLDERS AND THEIR EVALUATION OF THE RISKS AND OPPORTUNITIES OF CONNECTED AND AUTOMATED TRANSPORT

The first survey was conducted in autumn 2017 and aimed to:

- ascertain the participating experts' level of knowledge and their respective sources,

- obtain a nuanced picture of different CAM use cases, and

- find out how urban and mobility planners evaluated the opportunities and risks frequently mentioned in the general discourse.

First, participants were asked where they access information about the subject and/or how much first-hand experience they have with CAT. Just below 75% of the experts surveyed responded that they source information from the (specialist) media. Moreover, around 60% of those surveyed stated that they had already undertaken in-depth scientific research or been actively involved in planning or research activities. A considerably smaller group stated that they were involved in specific tests using CAVs (approx. 27%) or already had first-hand experience with CAVs (approx. 29%). Those in the latter three groups were also asked to specify which specific CAV use cases they had experienced (see Table 3.4.1).

When asked about the relevance of CAVs for their respective professional field, 68% of the participants believed CAVs had a high or very high potential to help develop innovative products or planning strategies. Respondents also believed that working with CAVs offered a high or very high potential to elevate the status of their own institution (approx. 38%) and to develop solutions that better meet the demands of customers. 78% of respondents felt that CAVs posed no threat to their professional field.

With regard to previous experience with CAVs, by far the most frequently encountered vehicles were automated shuttle buses (approx. 74%), which were well ahead of Level 4 cars (approx. 40%). At present, other use cases appear to play a less important role. A majority of respondents think that the different use cases will have a positive impact on their professional field: 61% of participants stated that automated shuttles would offer solutions to challenges and unresolved issues in their professional field (Figure 3.4.1).

All participants were asked to state how likely they thought automated modes of transport were to replace traditional forms of mobility. Respondents believed that

Table 3.4.1: **Respondents' first-hand experience and need (broken down by use case)**

	Which connected and automated vehicle applications have you experienced first-hand? (n = 149)			Which connected and automated vehicle applications do you expect to make a positive contribution within the context of your profession? (n = 193)		
	Responses		% of cases*	Responses		% of cases*
	N	%		N	%	
SHUTTLE	110	29.7	73.8	118	17.6	61.1
PUBLIC TRANSPORT	47	12.7	31.5	112	16.7	58.0
CAR SHARING	41	11.1	27.5	102	15.2	52.8
RIDE SHARING	37	10.0	24.8	86	12.8	44.6
LEVEL 5³	28	7.6	18.8	78	11.6	40.4
LEVEL 4⁴	59	15.9	39.6	62	9.3	32.1
FREIGHT TRANSPORT	33	8.9	22.1	62	9.3	32.1
OTHER USE CASE	15	4.1	10.1	7	1.0	3.6

* Multiple responses were possible, resulting in values > 100
Source: AVENUE21

connected and automated sharing services had the highest potential for crowding out other transport options. 97.6% of those surveyed expect that this use case will displace one of the existing forms of transport, closely followed by connected and automated private cars (96.2%) and driverless public transport (93.4%). Respondents thought fully automated freight transport would have a considerably lower impact (49.3%; see Fig. 3.4.2).

When the question was reversed, experts presume that traditional public transport will be most heavily impacted by the displacement effect (93.4% of experts believe that at least one of the automated forms of mobility mentioned in the survey will displace traditional public transport), followed by conventional private cars (89.6% of respondents). Each of these high figures underscores the often-cited argument that a core characteristic of automation will be the blurring of boundaries between individual and public transport ("hybridization", Lenz/Fraedrich 2015: 185).

Although considerably fewer specialists expect that cycling (61.1%) and walking (56.4%) will become less relevant, they do think that even these active forms of mobility are highly likely to be displaced by CAT, which stands in direct contrast to the objectives of the existing transport policy, i.e. to encourage active mobility.

One other relevant point is the high potential CAVs have for displacing traditional forms of transport (see Fig. 3.4.2). Here respondents gave a considerably higher rating for fully automated freight transport compared to conventional private cars (42.7%). This means that the process of automation will not only affect people's transport habits, it also raises the question of which journeys could be completely delegated to machines.

To find out how respondents view frequently raised arguments on the potential impacts of CAVs, these claims were presented as possible advantages and disadvantages – shown here in two separate tables (Tables 3.4.2 and 3.4.3) – for participants to evaluate.

The 14 advantages presented in the survey were grouped based on a factor analysis, a statistical method used to group variables that correlate to one another (see Table 3.4.2, Backhaus et al. 2019). Using this method, we were able to ascertain two factors. The first factor combines structural policy and social aspects (with an explained variance of 28%). The second groups together those aspects that are more economically relevant (explained variance: 21%). If the individual statements are listed based on their average level of approval, the economically relevant advantages are almost always given higher approval ratings than the structural policy and socially relevant factors.

The structural policy and socially relevant benefits tend to receive less approval (see Table. 3.4.2): the mean value of responses even falls within the category of "slight rejection" (< 4.0) for the final two statements. The respondents thus tend to believe that CAT will not lead to a stabilization of rural areas or the freeing up of intra-urban areas. This last finding stands in striking contrast to the results of a number of studies and to urban planning actors' core judgement that CAT will allow intra-urban traffic areas to be freed up so that they can be repurposed (see Chaps. 1, 4.1 and 4.3).

Respondents consider the link between CAT and the production and utilization of digital data generated during operation to be particularly relevant (see Table 3.4.2). Among the advantages listed in the survey, the

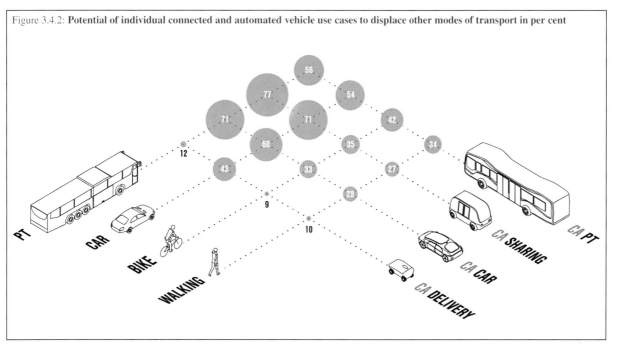

Figure 3.4.2: **Potential of individual connected and automated vehicle use cases to displace other modes of transport in per cent**

statement "Increasing automation and networking of transport will lead to the collection of larger volumes of additional data to be used for efficient control of transport" was calculated with an arithmetic mean[5] of 5.33 as the second most approved benefit.

However, respondents also considered the potential associated risks and counterproductive developments to be considerable (see Table 3.4.3). If we list the disadvantages according to their respective approval ratings, three of the four most strongly weighted disadvantages concern data security:

■ 91% agree with the statement that "Connected and automated transport will lead to large volumes of additional data being collected and used by third parties" (with an arithmetic mean of 6.00 – one of the highest values in our data set).

■ 78% state that "Connected and automated transport will lead to large volumes of additional data being collected and used for continuous monitoring" (with an arithmetic mean of 5.44).

■ 72% agree with the statement that "Connected and automated transport will lead to transport becoming a security risk due to hacking" (with an arithmetic mean of 5.13).

The impacts CAV will have structurally and in terms of social policy – be they negative or positive – are considered by respondents to be of little importance. Participants seemed to offer little approval of the statements concerning urban sprawl, the waning significance of brick-and-mortar retail, the risk posed to certain professions and the basic provision of spatial mobility (see Table 3.4.3). Moreover, respondents assume that the

Table 3.4.2: **Expert assessment of the opportunities arising from the introduction of connected and automated transport**

INCREASING AUTOMATION AND NETWORKING OF TRANSPORT WILL LEAD TO ...		STRUCTURAL POLICY AND SOCIAL FACTOR	ECONOMICALLY RELEVANT FACTOR
	Mean values	Factor loading	
... new solutions in the logistics sector.	5.67	0.199	0.639
... large volumes of additional data being collected and used for efficient control of transport.	5.33	0.160	0.734
... an increase in transport safety.	5.31	0.428	0.518
... strengthened intermodality through services provided over the last passenger transport mile.	5.25	0.690	0.297
... increased mobility comfort.	5.18	0.325	0.582
... increased sharing services.	4.89	0.542	0.354
... an increase in the performance capability of the transport network.	4.79	0.274	0.576
... public transport services becoming more cost-efficient.	4.71	0.567	0.494
... boosting of the economy.	4.52	0.049	0.523
... public transport services being expanded.	4.5	0.557	0.375
... creation of socially inclusive mobility services.	4.42	0.671	0.340
... decarbonization of the mobility system.	4.03	0.605	0.277
... stabilization of rural areas.	3.87	0.771	-0.011
... freeing up of intra-urban areas.	3.82	0.790	0.143
Factors: mean value[6]		4.44	5.26
Factors: explained variance		28%	21%

1 = disagree. 7 = totally agree
Source: AVENUE21

number of motor vehicles on roads will not fall but rise in the coming years.

Their responses suggest a belief that this increase will also not be short term: with a mean value of 3.06, participants reject the statement "An increase in traffic volumes due to automated driving is an effect of the transition phase" (see Fig. 3.4.3). At the same time, with a mean value of 5.26, respondents clearly agree with the statement "Higher traffic volumes should be counterbalanced through legislation".

It is notable that the respondents from German-speaking countries were far more clearly in agreement with these statements than those responding to the English survey. While the average approval figures for the German survey are significantly different for these two questions, for the English survey, the mean approval values stood at 4.62 (regulation) and 3.9 (increase temporary) and were thus so close that the differences are not significant at a CI of 95%.

Table 3.4.3: Expert assessment of the risks and impacts involved in the introduction of connected and automated transport

CONNECTED AND AUTOMATED TRANSPORT WILL LEAD TO ...	MEAN VALUES
... large volumes of additional data being collected and used by third parties.	6.00
... large volumes of additional data being collected and used for continuous monitoring.	5.44
... increased motor vehicle traffic volumes.	5.26
... transport becoming a security risk due to hacking.	5.13
... increasing urban sprawl as peripheral locations will become increasingly attractive.	4.89
... production and delivery chains being completely overhauled.	4.73
... the decreased significance of offline retail trade compared to e-commerce.	4.69
... a threat to jobs that, at first glance, do not have any direct connection to vehicle steering.	4.53
... privatization threatening the basic provision of spatial mobility.	4.08

1 = disagree, 7 = totally agree
Source: AVENUE21

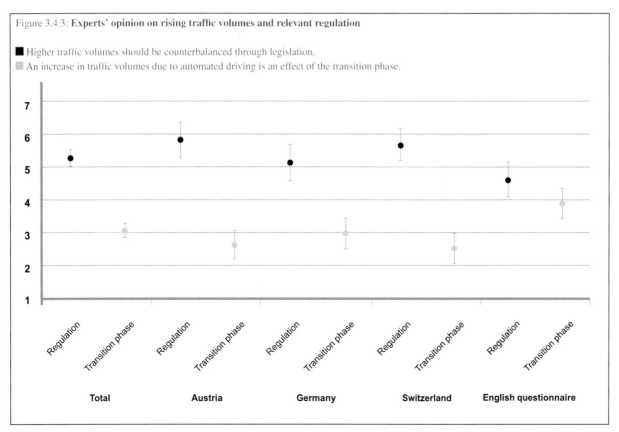

Figure 3.4.3: **Experts' opinion on rising traffic volumes and relevant regulation**

■ Higher traffic volumes should be counterbalanced through legislation.
■ An increase in traffic volumes due to automated driving is an effect of the transition phase.

| Total | Austria | Germany | Switzerland | English questionnaire |

Mean values and 95% confidence intervals: 1 = disagree, 7 = totally agree
Source: AVENUE21

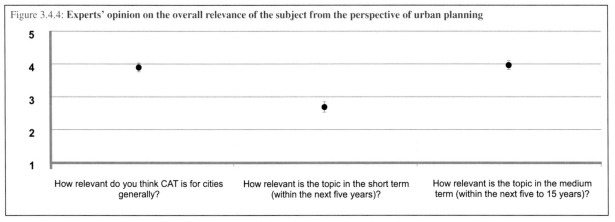

Figure 3.4.4: **Experts' opinion on the overall relevance of the subject from the perspective of urban planning**

How relevant do you think CAT is for cities generally?

How relevant is the topic in the short term (within the next five years)?

How relevant is the topic in the medium term (within the next five to 15 years)?

Mean values and 95% confidence intervals; 1 = not relevant, 5 = extremely relevant
Source: AVENUE21

3.4.3 SECOND SURVEY: STAKEHOLDERS AND SCOPE FOR ACTION IN TOWNS AND CITIES DURING THE INTRODUCTION OF CONNECTED AND AUTO-MATED VEHICLES

We carried out a second survey in the winter of 2018/2019. 216 experts agreed to take part, 98 of whom had already participated in the first survey.

The second survey focused on:

- potential actions and planning options in urban regions,

- how stakeholders' professional backgrounds influenced their evaluations,

- potential for cooperation and conflict, and

- examples of measures that could influence the effects of CAT.

Participants were thus asked to judge how relevant CAT would be over certain periods of time. Here it became evident that the respondents only considered the topic to be of any real importance within the medium term (see Fig. 3.4.4); in their view, CAT is unlikely to be relevant over the coming five years.

Furthermore, participants were asked to evaluate the suitability of a range of settlement structures for the use of CAVs (see Fig. 3.4.5). Industrial and commercial areas were believed to be most suitable, with suburban settlement areas coming second, closely followed by new residential neighbourhoods. However, the mean ranks of both settlement types are so close in value that it cannot be determined (at a statistical certainty of 95%) which of the two should be ranked second and which third. What is clearer, however, is the ranking of the remaining types of settlement: post-war urban neighbourhoods are clearly ranked fourth and (edges of) historical town and city centres undoubtedly come in last. The evaluation given by experts in this survey evidently mirrors AVENUE21's analysis on automated drivability (see Chap. 4.4).

Subsequently, a series of questions were posed which aimed to ascertain the potential for conflict between urban areas and various stakeholders based on their diverging interests (see Fig. 3.4.6). The potential for conflict between urban areas and international market

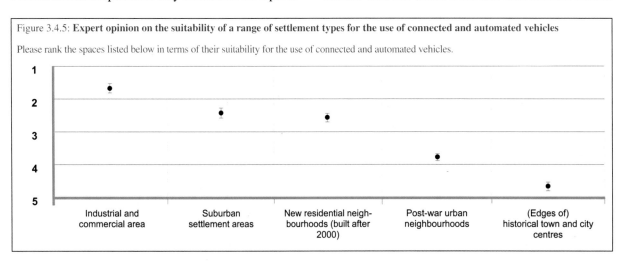

Figure 3.4.5: **Expert opinion on the suitability of a range of settlement types for the use of connected and automated vehicles**

Please rank the spaces listed below in terms of their suitability for the use of connected and automated vehicles.

Industrial and commercial area

Suburban settlement areas

New residential neigh-bourhoods (built after 2000)

Post-war urban neighbourhoods

(Edges of) historical town and city centres

Mean ranks and 95% confidence intervals; 1 = best suited, 5 = worst suited
Source: AVENUE21

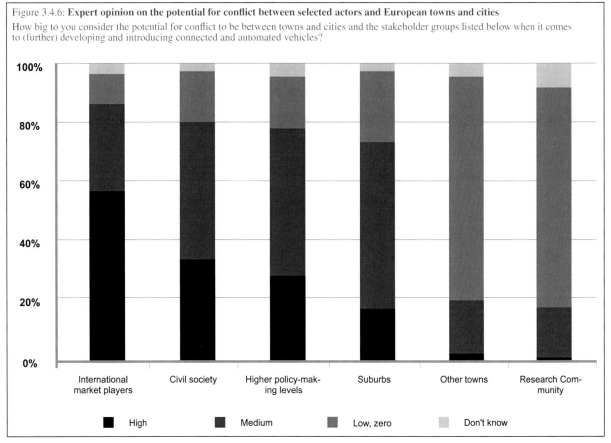

Figure 3.4.6: **Expert opinion on the potential for conflict between selected actors and European towns and cities**
How big to you consider the potential for conflict to be between towns and cities and the stakeholder groups listed below when it comes to (further) developing and introducing connected and automated vehicles?

Source: AVENUE21

players was considered to be the highest at just under 60%. A slightly lower percentage believed there was the likelihood of conflict between cities and towns and civil society, higher policymaking levels and the rural-urban fringe (between approx. 20% and approx. 30%). A majority (around 75%) of respondents feel there is no or only a minimal potential for conflict between the urban areas themselves as well as between cities and towns and researchers.

One aim of the second survey was to gather data on experts' opinions regarding which of the many approaches towns and cities should take to respond to the introduc-

tion of CAVs (see Fig. 3.4.7). Respondents were offered a range of answers, each linking a time component (taking action from the start vs. waiting to see how things start to develop) with a type of action (encouraging CAT vs. CAT regulation). The results shown below illustrate that experts feel it is more important to support CAT than to regulate the technology: the statement "Providing active support from the outset during introduction" is seen as the most important, followed by "Awaiting the first developments and intervening in a supporting manner, if needed". Respondents named "Await the first developments and intervene in a guiding manner, if necessary" and "Actively intervene in the introduction process by

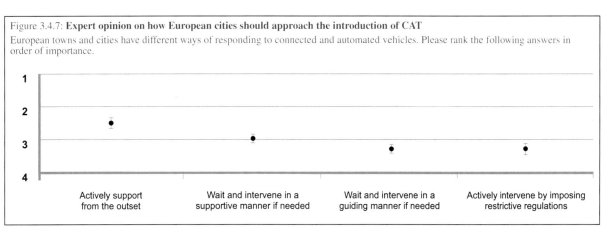

Figure 3.4.7: **Expert opinion on how European cities should approach the introduction of CAT**
European towns and cities have different ways of responding to connected and automated vehicles. Please rank the following answers in order of importance.

Mean ranks and 95% confidence interval; 1 = most important, 4 = least important
Source: AVENUE21

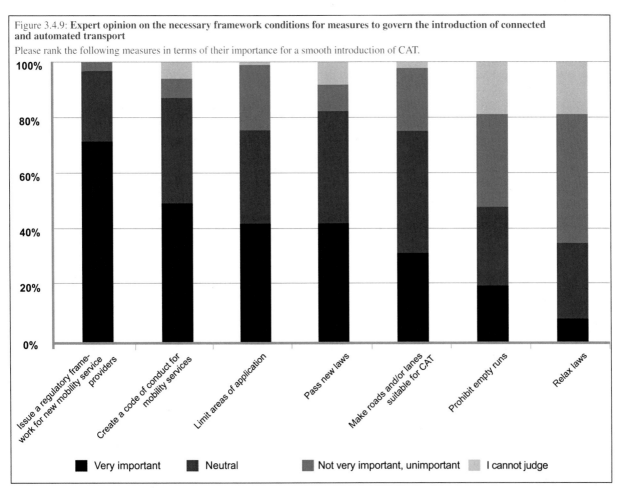

Figure 3.4.8: **Expert opinion on the need for measures that cities and towns can take to prepare for the introduction of connected and automated transport**

How urgent do you think the following steps are when preparing for the introduction of connected and automated transport?

Develop a design concept for future mobility (Education)

Set regulatory frameworks (Enforcement)

Integrate CAT into existing mobility concepts and missions (Enforcement)

Involve companies (Economy)

Inform the public about opportunities and risks (Education)

Build expertise, e.g. in administrative authorities (Education)

Assign responsibilities at an administrative level (Enforcement)

Conduct tests (Engineering)

Connect with other towns, cities and regions (Education)

Identify measures to prepare streetscapes (Enforcement)

Develop a catalogue of measures to prepare streetscapes (Enforcement)

Promote links between urban administrative authorities (Education)

Provide resources: human, financial, etc. (Economy)

Mean values and 95% confidence interval; 1 = not urgent. 5 = very urgent
Source: AVENUE21

Figure 3.4.9: **Expert opinion on the necessary framework conditions for measures to govern the introduction of connected and automated transport**

Please rank the following measures in terms of their importance for a smooth introduction of CAT.

Issue a regulatory framework for new mobility service providers

Create a code of conduct for mobility services

Limit areas of application

Pass new laws

Make roads and/or lanes suitable for CAT

Prohibit empty runs

Relax laws

■ Very important   ■ Neutral   ■ Not very important, unimportant   ▨ I cannot judge

Source: AVENUE21

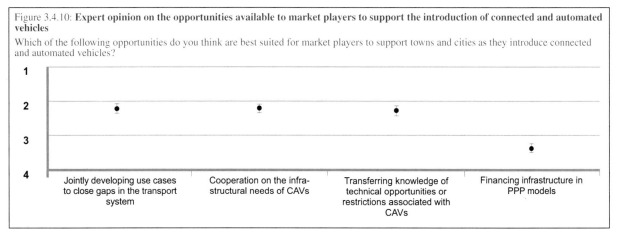

Figure 3.4.10: **Expert opinion on the opportunities available to market players to support the introduction of connected and automated vehicles**

Which of the following opportunities do you think are best suited for market players to support towns and cities as they introduce connected and automated vehicles?

Mean ranks and 95% confidence interval; 1 = most important, 4 = least important
Source: AVENUE21

imposing restrictive regulation" as the third- and fourth-best options, with an almost equal level of approval.

Furthermore, respondents were asked to state which of the measures available to towns and cities they deemed necessary (see Fig. 3.4.8). "Developing a vision for future mobility", "Setting of regulatory frameworks" and "Integrating CAT into existing mobility visions and/or missions" were considered to be particularly urgent (mean values of 4.2 or around 4); however, it should be noted that the mean values for the remaining items were all within a range of 3.4 to 3.75 and thus were almost identical.

This sheds light on a dilemma currently facing urban and mobility planners: the first survey confirmed that the issue of CAVs was highly relevant for professional groups involved in the broader field of urban and mobility planning; meanwhile, the second survey indicated that CAVs will be an urban development issue that will be highly relevant over the medium term (Figure 3.4.4). However, there does not appear to be consensus concerning the necessary measures that towns and cities should adopt (see Fig. 3.4.8). It is evident that negotiations will need to take place, but there are little to no clues as to how this should happen.

If towns and cities wish to manage the introduction of CAT, they can select measures from a range of different areas. The experts were asked how important they considered the areas of economy (e.g. financial incentives, pricing measures, fiscal policy), enforcement (e.g. statutory measures, regulatory policy), education (e.g. measures to raise awareness, communication, information) and engineering (e.g. technical execution of planning, road technology and supply-side measures) to be for the support and facilitation of CAT (see Fig. 3.4.8). Here the results show that the experts we surveyed considered economic measures (two aspects) as relatively neutral, with an arithmetic mean of 3.22 on a scale of 1 (not urgent) to 5 (very urgent). The measures that fall into the enforcement (mean value of 4.2: five aspects), engineering (mean value of 3.96: one aspect) and education (mean value 3.81: one aspect) categories were, however, seen to be (slightly) important.

In the next step, respondents were asked to assess the necessary frameworks for specific measures regarding the introduction of CAT (see Fig. 3.4.9). Respondents considered the most important to be "Draw up policies and principles for new mobility service providers" (approx. 70%), followed by the creation of a code of conduct for mobility service providers (just under 50%).

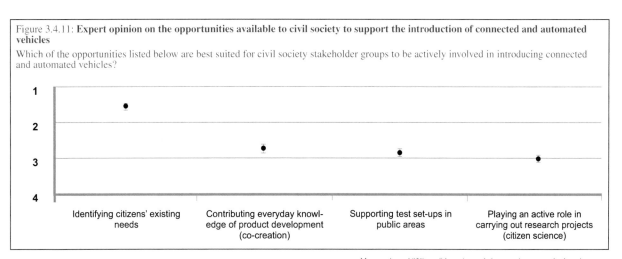

Figure 3.4.11: **Expert opinion on the opportunities available to civil society to support the introduction of connected and automated vehicles**

Which of the opportunities listed below are best suited for civil society stakeholder groups to be actively involved in introducing connected and automated vehicles?

Mean ranks and 95% confidence interval; 1 = most important, 4 = least important
Source: AVENUE21

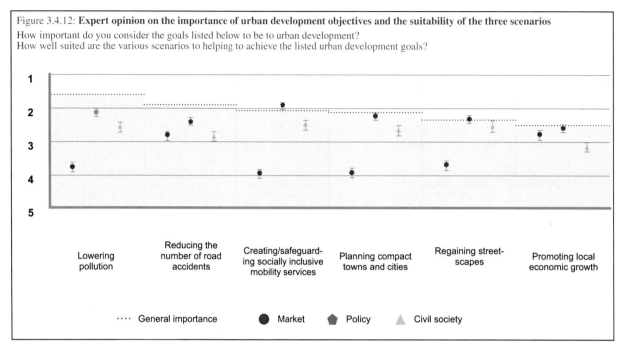

Figure 3.4.12: **Expert opinion on the importance of urban development objectives and the suitability of the three scenarios**
How important do you consider the goals listed below to be to urban development?
How well suited are the various scenarios to helping to achieve the listed urban development goals?

Urban development goals: 1 = extremely important, 5 = extremely unimportant; suitability: 1 = scenario is very well suited, 5 = scenario is not well suited at all
Source: AVENUE21

Prohibiting empty runs and relaxing laws were considered to be slightly less important. However, a proportionally high number of respondents (just under 20%) stated that they did not feel able to judge the importance of the measure.

Moreover, participants were asked for their opinion on how towns and cities could be supported in their efforts to implement CAVs by market players (see Fig. 3.4.10) and civil society stakeholder groups (see Fig. 3.4.11).

Here respondents were presented with a series of possible measures and asked to rank them from "most important" to "least important" depending on their suitability. If we compare those measures experts ranked in the middle of the list with regard to the steps the market can

adopt to support towns and cities during the introduction of CAT (see Fig. 3.4.10), respondents appear to be in agreement that "Financing infrastructure in PPP models" is the least important of the suggested possibilities. With regard to the other suggested measures ("Jointly developing use cases to close gaps in the transport system", "Working together on the infrastructural needs for CAVs" and "Transferring knowledge of technical opportunities and restrictions with CAVs"), respondents appear to show no clear preference when it comes to selecting the measures they consider most suitable.

A similar picture emerged when respondents were asked about the possibilities available to civil society stakeholders to support towns and cities (see Fig. 3.4.11). Here respondents are largely in agreement that the best

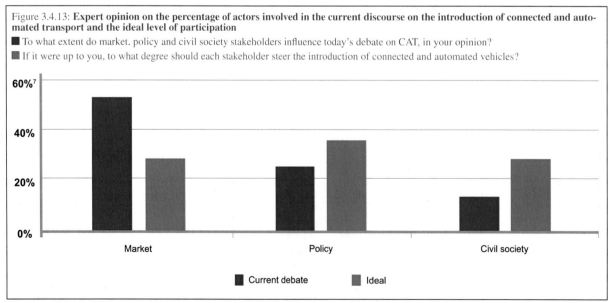

Figure 3.4.13: **Expert opinion on the percentage of actors involved in the current discourse on the introduction of connected and automated transport and the ideal level of participation**
■ To what extent do market, policy and civil society stakeholders influence today's debate on CAT, in your opinion?
■ If it were up to you, to what degree should each stakeholder steer the introduction of connected and automated vehicles?

Source: AVENUE21

way for civil society to be involved is to support towns and cities in their efforts to introduce CAVs by drawing attention to the existing mobility needs of residents based on their everyday experiences. However, with regard to the other statements ("Contributing everyday knowledge of product development", "Supporting test set-ups in public areas" and "Playing an active role in carrying out research projects"), the respondents are almost always divided on the ranking of importance.

During the AVENUE21 project, three scenarios were developed outlining the possible future effects the integration of CAT will have on European towns and cities if this introduction is primarily influenced by three different stakeholders (the market, policy, civil society; see Chap. 5). It was thus one of the aims of the second survey to underpin the scenarios developed by our research team with expert opinion. Respondents were therefore also asked how suitable they considered the three scenarios to be to achieving frequently stated urban development objectives (see Fig. 3.4.12). All the objectives listed in the survey were generally considered to be relevant. Respondents felt the most important goal was to alleviate the burden on the local and global environment and climate, followed by reducing the number of road accidents and fatalities. The aim of promoting local economic growth/location policies, on the other hand, was deemed the least important. When different scenarios were compared, the market-driven approach was almost always given the worst score or was always judged to be less important than the two other scenarios. It only ranked ahead of the civil society-driven scenario with regard to the goal of promoting local economic growth and received a similar score when it came to reducing road accidents and fatalities. Of the three scenarios, the policy-led scenario was consistently given the best ranking. This illustrates that European experts are undoubtedly sceptical about the free market's ability to

effectively manage the introduction of CAT, and instead believe primarily in the capacity of policymakers and planning authorities and/or the towns or cities to develop solutions.

In order to better evaluate how relevant these scenarios are to the real world, respondents were asked to state which stakeholder groups are most involved in the current debate on CAT (see Fig. 3.4.13). Respondents stated that the market has a powerful influence (over 50%), and ranked quite far ahead of policy (approx. 25%) and civil society stakeholders (approx. 15%). If the desired balance of power and involvement were in place, market, policy and civil society stakeholders would have a relatively equal say in how CAVs were introduced, with policymakers having slightly more control (approx. 37%) than the market and civil society (approx. 30% each). There is therefore a clear mismatch between the current discourse and the positive impact each actor is seen to potentially be able to bring to the table.

3.4.4 SUMMARY

The aim of conducting these two surveys was to ascertain the opinions of experts involved in a range of professions linked to urban and mobility planning with regard to CAVs in general and current urban development goals in particular. It was the opinion of the over 300 specialists who participated in the two surveys that a debate on the issue of CAVs should and must take place, not only in their respective professional fields but also more generally within the wider context of urban development. One key outcome of the surveys is that the results paint a nuanced picture that oscillates between scepticism and hope. For example, the 211 respondents who participated in the first wave believe (see Table 3.4.2) that CAT will result in:

Figure 3.4.14: **The surveys reveal a new urban mobility paradigm**

A NEW MOBILITY PARADIGM ...

WHICH MODE OF TRANSPORT SHOULD I CHOOSE FOR MY JOURNEY?

SHOULD I DRIVE MYSELF OR LEAVE IT TO A MACHINE?

... IN SOME AREAS WITHIN URBAN REGIONS

- new solutions in the logistics sector,

- large volumes of additional data being collected and used for efficient control of transport,

- improved road safety,

- strengthened intermodality through services provided over the last passenger transport mile, and

- increased mobility comfort.

However, they appear less convinced that CAT will enable

- intra-urban areas to be freed up,

- rural areas to be stabilized,

- the mobility system to be decarbonized, and

- socially inclusive mobility services to be created.

Data generated as a result of connected and automated processes are viewed with notable ambivalence; the greatest scepticism is shown towards the following risks and consequences of CAT (see Table 3.4.3):

- collected data will be used by third parties,

- collected data will be used for continuous monitoring, and

- transport could become a security risk due to hacking.

According to the survey, another problem (see Table 3.4.3) is the fact that CAVs will lead to the volume of traffic increasing instead of decreasing (a number of other scenario-based studies have also reached the same conclusion; see Chap. 4.3). As it is assumed that an increase in traffic will not simply be a temporary effect, respondents in the first survey expect that regulation will be necessary to avoid the expected growth in vehicle numbers (see Fig. 3.4.3). Respondents who participated in the first survey suspect that CAVs and the new business models based on this nascent technology are highly likely to replace other services (see Fig. 3.4.2). This was viewed

- as clearly having a positive impact because sharing services based on connected and automated mobility will reduce the volume of traditional private cars (71%),

- as equally positive because public transport services based on connected and automated mobility will replace traditional public transport (56%) and traditional private cars (54%),

- as moderately positive because a traditional private car will be replaced by a "smarter" version (60%), which, however, will also lead to continued use of motor vehicles as a form of transport.

Another factor considered problematic (although this was a concern for far fewer respondents) is that CAVs may dampen enthusiasm for active forms of mobility, which are currently being heavily promoted in European cities. The biggest competition arises from

- connected and automated public transport (42% likelihood of displacing cycling and 34% chance of replacing walking),

- connected and automated sharing (35% or 27%), and

- connected and automated private cars (33% or 28%).

When we asked respondents for their opinion on the possibility of other services being crowded out, the two key takeaways were:

- the service "hybridization" theory and thus the blurring of the boundaries between individual and public transport; there is evidence to support this argument but it is potentially too short-sighted as

- the shift will ultimately not only impact the choice of transport but will also present travellers with a new option: whether to complete the journey themselves or to delegate this task to a machine.

In the second survey, the main emphasis was placed on aspects that concerned the management and evaluation of the scenarios developed as part of the project. The 216 participants consider that the debate on CAT is mainly led by companies and/or the market (see Fig. 3.4.13), but believe that policymakers should play the leading role. They consider the most pressing need for action to be with regard to cities and/or policymakers and planning authorities (see Fig. 3.4.8), who they suggest should

- develop a concept for future CAT-based mobility,

- set regulatory frameworks,

- integrate CAT into existing mobility concepts,

- more heavily involve businesses, and

- inform the public about the opportunities and risks.

The experts we surveyed were not in agreement about the possible courses of action that can be taken by local/regional policymakers and town and city planning authorities. The two key views, which received an almost equal level of support in the survey, contradict each other (see Fig. 3.4.7): the restrictive regulation of CAVs and the proactive encouragement of their implementation both received almost the same level of approval.

Among the options given to steer CAT implementation, respondents felt that a set of policies and principles and/or a code of conduct for "new" mobility service providers were the most important. A general relaxing of laws or the prohibition of empty CAV runs, a measure frequently cited in the literature (Fagnant/Kockelman 2015), were seen to be less relevant (see Fig. 3.4.9).

Respondents feel the strengths of private sector companies would best serve cities if these actors are involved in cooperative projects (defining applications, infrastructure needs, allowing knowledge transfer), while they clearly reject the idea of involving market actors in infrastructure development and the financing of such measures through PPP models (see Fig. 3.4.10). Consequently, the experts believe the biggest potential for conflict to be between cities and international corporations, followed by conflict with civil society (see Fig. 3.4.6).

The experts are of the opinion that cities could benefit if civil society's needs and everyday knowledge are taken into consideration during the introduction of CAVs (see Fig. 3.4.11). The idea of civil society actors playing an active role as part of "citizen science" concepts was least popular among the experts.

Alongside issues that relate directly to transport, respondents were also asked to give their verdict on the importance of urban development goals (see Fig. 3.4.12). Here they believe it is vital to:

- reduce the environmental impact,

- bring down the number of road accidents,

- plan a compact city, and

- safeguard socially inclusive mobility services.

Moreover, respondents were also given a brief outline of the three scenarios developed as part of this project and asked to give their verdict on which of the scenarios would most likely meet the six outlined urban development goals. With regard to each goal, the policy-driven scenario was determined to have the most positive impact, followed by a civil-society driven approach. Respondents were only convinced the market-driven approach would be more successful than the civil society-based concept when it came to encouraging economic growth, but even in this regard, the experts still believed the policy-driven approach would be most effective.

The survey respondents still place a substantial amount of trust in the public authorities. However, it is clear that

- it is necessary to address the social challenges that will arise through the introduction of CAT in good time;

- policymakers and planning authorities and/or towns and cities should act consistently – whether they should take a proactive or reactive approach remains subject to debate;

- it remains unclear which measures should be applied as part of this approach, and

- CAT is far from likely to meet the highly optimistic expectations.

When and how policymakers and urban planners should exert a controlling influence will be decisive in determining whether the positive impacts will outweigh the negative, and whether those unwanted impacts can be avoided altogether.

---

1 Julia Dorner played a vital role in designing and carrying out not only the two surveys but also their statistical analysis and the interpretation of the results.

2 First survey: 149 fully completed German questionnaires, 62 English questionnaires.
Second survey: 181 fully completed German questionnaires, 35 English questionnaires.

3/4 No automated Level 5 driving systems existed (in accordance with SAE J3016) at the time the survey was conducted. Nonetheless, 28 of those surveyed stated that they had experience with Level 5 vehicles.

5 It is frequently debated whether Likert scales (i.e. ordinal scales) can be calculated as a metric (interval) scale. Due to the fact that parametric analysis methods have been proven effective at preventing various violations of the statistical criteria (see, for instance, Norman 2010) and, at the same time, open up considerably more possibilities for analysis, we decided to use Likert scales and Likert items on a metric scale.

6 Despite the seemingly obvious loading for the economic factor, one item ("Increasing automation and networking of transport will lead to boosting of the economy") was not included in the economic factor based on a reliability analysis.

7 An additional category ("other") was mentioned but only some participants filled this section in, resulting in percentage values that added up to less than 100%. Those most frequently named in this category were researchers.

Figure 3.5.1: **Population density in the studied regions in 2011, inhabitants per km²**

LONDON

RANDSTAD

VIENNA

Image: AVENUE21; source: Eurostat 2011

Figure 3.5.2: **Population density in Europe in 2014 on a 10 x 10 kilometre grid**

3.5

DEVELOPMENT OF TRANSPORT AND SETTLEMENT POLICY: LONDON, RANDSTAD, VIENNA

To provide case studies that cover the full spectrum of the "European city" and the effects that CAT could have in these areas, we selected three regions for analysis: the region of Greater London, the region of Randstad in the Netherlands and Vienna/Lower Austria. By analysing these regions, we hope not only to focus attention on the circumstances and urban challenges specific to each area but also to highlight the key role context plays with regard to the implementation of CAT and to suggest various options for policy and planning action. What the future holds for these regions depends not just on the spatial conditions and infrastructures, but, crucially, on the approach – now and in years to come – that will be adopted by policymakers and planning authorities (see Chaps. 3.2 and 4.6).

Analysing these specific towns and cities thus allows us to illustrate the wide range of possible future applications of CAT in respect of existing spatial structures, urban planning concepts and perceived challenges. In Chapter 4.5, we expand our research to also examine pioneering schemes globally.

3.5.1 METHODOLOGY AND SELECTION

This analysis focuses on the ways in which governance, mobility and urban development interact in selected localities/local contexts. The cities or regions analysed here (London, Randstad and Vienna) were selected based on a theory-led approach. The criteria for selec-

Figure 3.5.3: **Overview of the analysed regions**

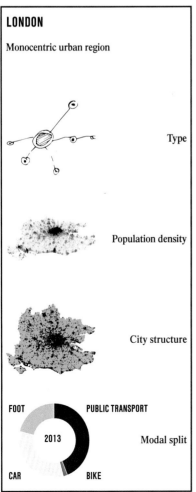
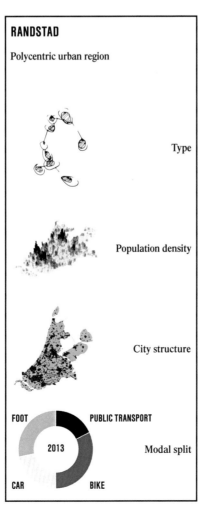
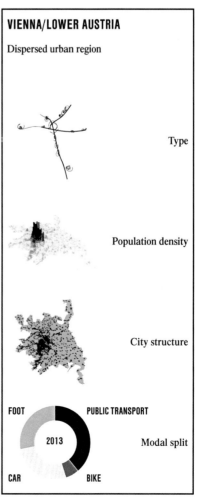

Source: AVENUE21, Eurostat (2017)

tion were primarily settlement structure requirements and the type of urban region, mobility cultures and infrastructures as well as planning and governance systems, which differ considerably in the three regions under analysis. Furthermore, the selected cities/regions are all in Europe and are characterized by dynamic economic and demographic development. Figure 3.5.4 compares the regions' transport and mobility policy measures, drawing upon the urban transport policy development path set out by Jones (2017) and mentioned here in Chapter 3.2.

Moreover, the cities as well as the urban regions are each considered a prime example of their respective settlement structure. In terms of transport infrastructure, it is significant that the city of Vienna is based on a predominantly concentric city model, the urban region of Randstad is mainly characterized by polycentric links (a linear city surrounding a "Green Heart") and London as well as its urban region (surrounded by a "green belt") has good transport links between the all-powerful centre and its surrounding satellite towns.

Figure 3.5.4: **Overview of transport and mobility policy in the three selected regions**

	STAGE 1 Accommodating traffic growth	STAGE 2 Encouraging modal shift	STAGE 3 Promoting liveable cities
GREATER LONDON	• Plan to build four concentric ring roads and radial roads • Dismantling of the tram system • Construction of urban expressways • One-way roads and parking spaces	• "Homes before Roads" initiative • Zones and day tickets on public transport • Construction of the Jubilee Line • Limiting transport costs and uniform zone system on public transport • Construction of the Docklands Light Railway and Thameslink • Planning of the Thameslink network	• Expansion of the public transport network (esp. improvement of stations and door-to-door services) • Introduction of the "Oyster Card" • Promotion of walking and cycling (cycle lanes, "Cycle Superhighways") • "Healthy Streets": a new narrative for London
RANDSTAD	• Merging different rail operators to form one national provider • Provision of a widespread motorway network across the whole of the country • First national long-term spatial development strategy • New policy instruments adapted to political, geographic and social realities (Randstad, "Green Heart")	• Nationwide ticketing and pricing system for public transport • Priority for public transport at traffic lights • "Stop de Kindermoord" ("stop killing children") protests in Eindhoven (1973) • "Straßenspieltag" – temporary road closures so children can play outside (1986) • Policy to encourage less car use and more journeys by public transport • Restrictive policies to limit distance travelled using vehicles • "Compact city" as a planning model	• Structural model for infrastructure and regional development • "Randstad 2040" strategy • "OV-chipkaart" • "OV-fiets": linking cycling and train travel • Encouragement of multimodal transport, improved mobility hubs and transport information • Decentralization of national policy – greater focus on regions • PPP models for infrastructural improvements • Implementation of pricing model for roads based on kilometres travelled
VIENNA/LOWER AUSTRIA	• Closure of some tram lines that were replaced by municipal bus services • Number of kilometres passengers could travel by tram reduced by a third • Radial concentric growth	• Introduction of VOR, a transport authority for the eastern region (1984) • Increase in public transport mileage and network lines (e.g. underground expansion) • Use of low-floor trams • Linear-shaped city expansion along settlement axes	• STEP 2025: boost ecomobility, lower MPT as a percentage of modal split to 20% by 2025 • Construction of U2, U5 underground lines; expansion of urban railway • €365 ticket for public transport, "WienMobil" app to improve inter- and multimodality • Encourage walking and cycling (e.g. creation of pedestrian zones and shared spaces, mobility agencies)

Sources: Greater London: Jones (2017); Randstad: Reid (2017), OECD (2014), MOT (2017), Alpkokin (2012);
Vienna/Lower Austria: Schubert (1985), Békési (2005), City of Vienna (2017)

Based on these various conditions, the cities and regions set different priorities with regard to transport and settlement development. Figure 3.5.5 gives an example of the key themes addressed in each respective area.

Figure 3.5.5: **Key areas for urban and mobility planning in the selected regions**

GREATER LONDON
• Transport connections to overflow cities (last-mile solutions)
• Decentralization of the region – more polycentric approach

RANDSTAD
• Inter- and multimodality (with a high percentage of cyclists)
• Settlement development along multimodal transport hubs

VIENNA/LOWER AUSTRIA
• Creating links between the city and surrounding region
• Improving public transport

Source: AVENUE21

If we compare the European regions chosen, we see, on the one hand, quite substantial differences resulting from each country's and region's respective historical development and unique settlement characteristics (see Figs. 3.5.4 and 3.5.5). On the other hand, some similarities can also be seen. The impressive transport and communication links between the metropolises and their surrounding regions, all of which transcend the boundaries of each city, defy the term "European city" (see Chap. 3.2). This element requires new management models that do not end at the city's boundary lines but encompass interregional links and emphasize the relevance of specific key (international) hubs (SUMP; Backhaus et al. 2019, Wefering et al. 2014). Connected and automated transport will create and allow for new links and thus have a substantial impact on the spatial and transport situation as well as the character of the European city. We will thus provide an overview of select issues affecting urban and mobility planning in the three regions and explore current CAT projects in the regions in more detail. In doing so, it becomes clear that the way in which CAVs are examined should depend on the context and that the potential and opportunities for desirable changes can be brought about if CAT is purposefully developed and managed. This approach to (future) CAT development and management will by no means redefine urban planning, but will instead take its place within historically evolving spatial development strategies.

Figure 3.5.6: **Polycentric links in Randstad**

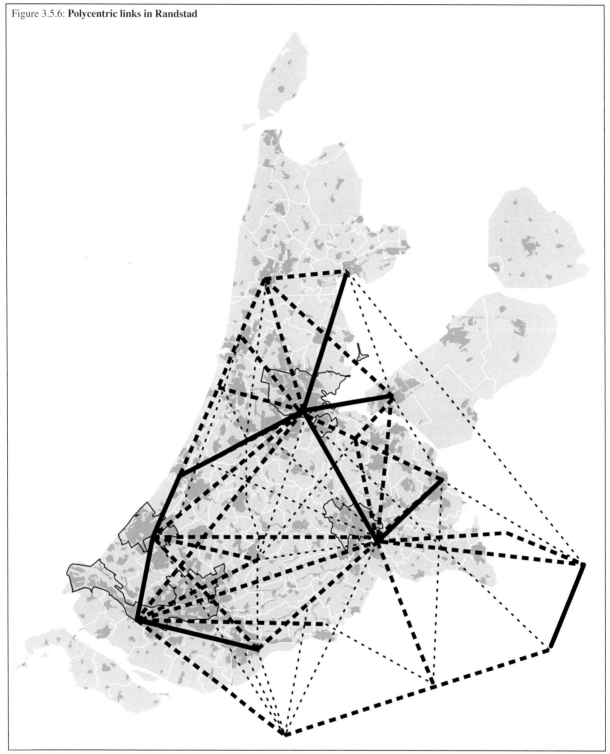

Source: AVENUE21

3.5.2 RANDSTAD

Randstad is a conurbation that is primarily characterized by a highly functional integration of urban zones and a polycentric settlement structure, which is also present in large parts of Europe (European Union 2011: 4). Polycentric urban structures are also one spatial development strategy deployed on the continent explicitly to bring about territorial cohesion (Hall/Pain 2006).

THE PATH TO AN INTEGRATED MOBILITY SYSTEM

The polycentric structure of the Randstad region, which consists of four large urban centres – Amsterdam, Rotterdam, The Hague and Utrecht – not only defines the settlement structure and development, but has long impacted the transport and mobility system. For many years, the transport planning model in the region was primarily shaped by the local interests of the individual cities, and this was particularly true of public transport. As recently as 2007, the OECD reported that Randstad does not have an integrated public transport system but fragmented systems and networks run by individual cities in the region (OECD 2007: 107). At the end of the last decade, however, increased efforts were made to develop an integrated mobility system in the Netherlands and thus in Randstad too. Policy papers such as "Mobiliteitaanpak" (2008) and the "Structural model for infrastructure and spatial planning" (2012) thus outlined the development of a coherent, integrated mobility system as a key transport and mobility objective. This was to be developed at the national level together with subnational authorities with the aim of ensuring national and regional mobility systems are more closely interlinked and more effectively aligned with one another. The various modes of transport should be better connected and the primary focus should be the promotion of multimodal transport and multimodal transport hubs (Ministerie van Infrastructuur en Milieu 2015: 10).

An important step towards the development of a coherent, integrated mobility system and, in particular, better linkage between public transport systems in the Randstad region came in the shape of the "OV-chipkaart", a scheme initiated by the Dutch government back in 2010 but only introduced nationwide (and beyond into Belgium and Germany) in 2012 (Ministry of Transport, Public Works and Water Management 2010: 5). The "OV-chipkaart" is a chip card that enables a single electronic payment system for the entire Dutch public transport system, i.e. it applies to all national, regional and local transport authorities as well as their respective fare schemes (Roland Berger 2016: 31).

INTER- AND MULTIMODALITY AS WELL AS IMPROVEMENT OF TRANSPORT HUBS

Further initiatives designed to develop an integrated mobility system are focused primarily on connecting public transport and/or rail services with cycling, a mode of transport far more frequently used in the Netherlands than in other European countries. One example of such an initiative is the growing construction and improvement of bike parking racks and storage areas at train stations (Godefrooij 2012: 40). A nationwide bike-lending scheme – "OV-fiets" – was also introduced back in 2003 and has since been taken over by the Dutch railway operator. The scheme offers many stations, whose numbers have grown considerably in recent years – especially at train stations and particularly in the Randstad region – and is expressly designed to offer a solution for the final section of a train journey (Ministerie van Verkeer en Waterstaat 2009: 48).

The Randstad region is characterized by a very dense railway network with a high number of stations, and this factor has also received a growing level of attention throughout the course of settlement development

projects (Stead/Meijers 2015: 12). For instance, as part of the 2040 policy plan for the northern part of Randstad ("Structuurvisie Noord-Holland 2040"), a "transit-oriented development" has been outlined that aims to increase use of the surrounding areas and/or catchment areas of train stations for settlement development as well as other urban functions (Deltametropol 2013: 228). This will be achieved through spatial measures and schemes as part of a coordinated location policy (Provincie Noord-Holland 2015: 46). In the Randstad region, transport hubs are usually more than just transit points; they are also places where urban activities take place, and travellers arrive, live and work (Deltametropol 2013: 85).

Curtis and Scheuer (2016) summarize the unique features of this planning approach as follows: "However, during the decade since, it has become clearer that the dichotomy of public transport versus car does not need to be regarded in competition. Instead, it can be viewed as an opportunity to work towards intelligent solutions of task-sharing and mutual support between these modes, and for walking and cycling and the growing range of hybrid forms of transport that do not neatly fit the traditional categories of collective and individual such as shared cars and bicycles, online ride-sharing or user-responsive public transport services. This type of thinking around multimodal accessibility, rather than single-mode market shares can be understood as the most significant contribution to global practice in integrated transport and land use planning to emerge from the Randstad and its unique interplay of settlement patterns and transport networks" (Curtis/Scheurer 2016: 287).

CONNECTED AND AUTOMATED TRANSPORT IN THE NETHERLANDS AND RANDSTAD REGION

Within the context of the development of CAVs, it is important to consider the key objective of an integrated mobility system in the Randstad region and examine how this new technology could create opportunities to further integrate transport systems (to develop one single service) and thus also lead to greater connectivity within the region itself and, based on these developments, thus consider the possibilities CAT may offer with regard to linking up urban peripheries or peri-urban landscapes in the region.

According to a number of international comparisons, the Netherlands have some of the best test and development conditions for CAVs (KPMG 2018, Welch/Behrmann 2018). The activities in the Netherlands are based both on tests involving vehicles and the development of infrastructure (e.g. the expansion of high-speed mobile data transmission) as well as on the proactive design of policy strategies ("Declaration of Amsterdam 2016"). Another relevant example is the "WEpods" project. As part of this project, two self-driving e-shuttles (EasyMile EZ10) have been in place since 2015 in Gelderland, the province bordering Randstad to the east, and in the towns of Ede and Wageningen, and were explicitly tested as potential last-mile solutions, i.e. possible ways to complete the last mile either from or to a train station. These could represent a very cost-effective form of public transport – available around the clock and on demand – especially in areas with low demand, and also effectively help boost public transport integration (Scheltes 2018, Fig. 3.5.3).

When the tests were initiated, even Melanie Schultz van Haegen, Dutch Minister of Infrastructure and the Environment at the time, emphasized the potential CAVs held for creating a more flexible and integrated public transport system: "With the WEpod, we are entering a completely new stage of the voyage of discovery that the Netherlands embarked on with the aim of making transport more flexible, safer and cleaner" (Wageningen University & Research 2016).

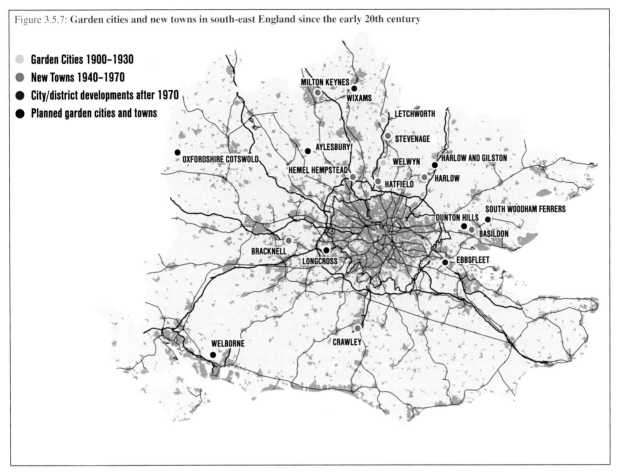

Figure 3.5.7: **Garden cities and new towns in south-east England since the early 20th century**

Garden Cities 1900–1930
New Towns 1940–1970
City/district developments after 1970
Planned garden cities and towns

MILTON KEYNES
WIXAMS
LETCHWORTH
STEVENAGE
AYLESBURY
OXFORDSHIRE COTSWOLD
WELWYN
HARLOW AND GILSTON
HEMEL HEMPSTEAD
HATFIELD
HARLOW
SOUTH WOODHAM FERRERS
DUNTON HILLS
BASILDON
BRACKNELL
LONGCROSS
EBBSFLEET
WELBORNE
CRAWLEY

Source: AVENUE21

3.5.3 GREATER LONDON

For more than a century, the growth of London has been intricately linked to developments in mobility and technology. Over the years, a series of satellite towns were built around the city's green belt, initially based on rail transport and, later, on personal mobility. Like other European regions that are witnessing the ongoing march of urbanization, the rapidly growing British capital faces considerable challenges in terms of settlement and transport development. However, London's stringent, historically developed, top-down planning approach to managing growth is unique.

THE HISTORY OF ENGLISH GARDEN CITIES AND NEW TOWNS

Even during the planning and development of London's first new towns at the start of the 20th century, the transport system that existed in the area surrounding the capital played a crucial role. Newly created towns such as Letchworth and Welwyn Garden City in the north of London were developed in 1903 and 1920 respectively and designed based on the concept of the garden city.

They were built along railway lines and planners made the town centres accessible on foot or by bike. These garden cities were designed to be small enough that pedestrians and cyclists could easily reach any point within the town in just 15 minutes; the railway provided access to the capital (Schmitz 2001: 48–49).

By the mid-20th century, the New Towns Act was passed, which allowed the construction of more new towns around London. Unlike the first developments built to accommodate the overspill of population from London, these new towns were designed purely to suit personal car mobility. A prime example of one of these developments is Milton Keynes situated to the north-west of London: it was the last new town to be built and was created in 1976.

THE CONCEPT TODAY

Even now some are pushing for new garden cities to be built around the capital (and in other parts of the UK) to ease the pressure on local services. In 2014, the Department for Communities and Local Government announced that a new garden city would be built in Ebbsfleet to the east of London to accommodate 15,000 residents (Department for Communities and Local Government 2015). In 2016, the department published guidelines whereby councils could submit an application to be chosen as a site for a new garden city. Subsequently, it was announced in early 2017 that a total of three new garden towns and 14 new garden villages would be built across the whole of England, with several of the new developments planned for areas close to London (Department for Communities and Local Government 2017 and Fig. 3.5.7).

The most pressing challenges facing London's urban planners and developers are the rapidly growing population in the region and the need to control this growth through decentralization. London's long-term urban development strategy has two pillars: on the one hand, the decentralization of growth ("Building the Polycentric City"; NLA 2017), and residents' health and well-being ("Healthy Streets for London"; TfL 2017) on the other.

Policy papers on the future development of transport and mobility in London and the Greater London region – such as the "London Infrastructure Plan 2050" – stress that settlement expansion (outside of London) and thus plans to build new overspill towns should be concentrated around existing, expanded or new transport corridors and stations (especially railway lines; Mayor of London 2014: 45). While no planning progress has been made on the garden villages of Longcross and Dunton Hills or the garden towns of Aylesbury and Harlow-Gilston besides choosing a location, initial development plans are already in place for the garden city of Ebbsfleet, especially in terms of transport and mobility.

TRANSPORT LINKS GENERATED THROUGH CONNECTED AND AUTOMATED VEHICLES IN CITY DISTRICTS AND ALONG "CONNECTED CORRIDORS"

Within this context, it thus comes as no surprise that Milton Keynes, i.e. a new town designed solely around personalized mobility, has hosted a series of CAT trials: initially it was part of the "LUTZ Pathfinder" research project and it is currently involved in the "UK Autodrive" research project, which is testing automated pods as part of the "Transport Systems Catapult" initiative (TSC 2017). As part of this, considerable funds, primarily provided by the Department for Transport, have been invested in research on these "Low-Speed Autonomous Transport Systems – L-SATS" which are explicitly seen as potential solutions for the last mile of urban mobility (TSC 2014: 2). The town of Milton Keynes is also convinced of the potential of CAT and plans to use it. As early as 2011, the city wrote in its transport strategy ("A Transport Vision and Strategy for Milton Keynes") that in terms of its public transport services, personalized public transport, such as automated pods, would be ideal in the long term for the city's grid road layout (Milton Keynes Council 2011: 42–43). Moreover, in 2015, the town refused to grant permission for the construction of a tram system. When stating its reasons, it made reference not only to cost but to the use

of pods as a public transport system (Smith 2015). This shows that plans currently being drawn up for many new garden cities – depending on the outcome of the tests in Milton Keynes – could already be designed to accommodate a public transport system that includes these more cost-effective pods.

If we turn back to Ebbsfleet Garden City, a site which is close to the A2 major road – and connected to a national and international (Eurostar) high-speed rail link – another aspect becomes relevant to debates concerning CAM: the use of CAVs on motorways. The Department for Transport, for instance, considers "the creation of connected corridors – initially to test, and then deploy, the technology – as a cornerstone of the UK [in the context of connected and autonomous vehicles]" (Hanson 2015: 4). One such test corridor is already in place on the A2/M2 motorway between London and Dover (Hanson 2015: 5). In collaboration with the government company in charge of the country's major roads and motorways (Highways England), the connection possibilities between vehicles as well as between vehicles and the infrastructure are being tested here against the backdrop of CAM (TRL Limited 2016: 1). There are also plans to carry out test runs involving CAVs – both here and on other stretches of motorway – in a subsequent stage.

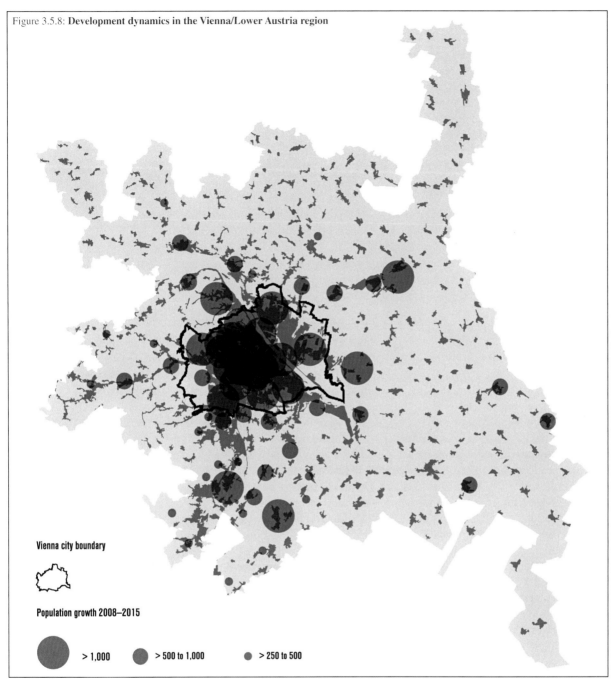

Figure 3.5.8: **Development dynamics in the Vienna/Lower Austria region**

Vienna city boundary

Population growth 2008–2015

> 1,000   > 500 to 1,000   > 250 to 500

Source: AVENUE21

3.5.4 VIENNA/LOWER AUSTRIA

Vienna is the epitome of a traditional European city that has long been influenced by a concentric design (Schubert 1985: 521). Since the late 1960s, the city has grown far beyond its established boundaries, becoming a dispersed urban region. Most notably in recent years, the population within Vienna and in its surrounding area has also grown considerably. This has further increased connectedness and integration within the overall metropolitan region. Administrative boundaries may exist but now have an ever-diminishing impact on everyday dealings and functional relationships (MA 18 2014: 88); however, they continue to be very present in governance and administrative structures. This poses particular challenges for management, especially in terms of spatial planning, transport

planning and location development. This can particularly be seen in the city/region's public transport system.

NECESSARY ACTION TO TACKLE THE CHALLENGES FACING THE METROPOLITAN REGION

As part of the current city development plan ("STEP 2025"), efforts are being made to harness the growth dynamic to benefit the population. Within this framework, a regional model with regional development axes was developed that focused heavily on regional links within the metropolitan region of Vienna (Vienna and parts of Burgenland and Lower Austria) and the Centrope region (which also includes western Hungary, western Slovakia and southern Czechia) as well as links within the metropolitan region (MA 18 2014: 91).

This model and the regional development axes aimed to ensure that population growth and suburbanization – processes that span administrative boundaries – went hand in hand with a managed settlement plan that was based on development and public transport axes (Dangschat/Hamedinger 2009: 108; Scheuvens et al. 2016). However, in future, this regional model for the entire metropolitan region will need to be given greater consideration within the various public authorities. A key factor here are city-regional governance structures that complement the administrative structures of public authorities, which have evolved over time and are usually designed to represent local interests, by giving a voice to various city-regional interests, such as the city-regional model. Such structures could make it possible to tap existing potential for cooperation more effectively, especially in areas such as regional planning, transport planning and location development (MA 18 2014: 91).

THE IMPORTANCE OF SETTLEMENT AND TRANSPORT DEVELOPMENT IN THE METROPOLITAN REGION

One element that is crucial here is the interlinking of settlement and transport development. However, it will not only be a matter of conducting coordinated settlement development along public transport axes. Instead, it will be necessary to also improve public transport services on a city-regional scale: while public transport has always played a vital role within the Austrian capital's urban transport system and accounts for one of the highest modal splits across Europe's capital cities (39% in 2016), if we compare these figures with those for the city-region and surrounding areas, i.e. among commuters, we see a considerably lower modal split share of 21% (City of Vienna 2014: 103). In the future, however, urban development plans and mobility and transport services across the entire metropolitan region – irrespective of administrative boundaries and various competencies – will need to be considered as an integrated system. This will require, first and foremost, effective regional cooperation at various transport and settlement policy levels as well as concepts for a controlled settlement development along public transport axes, in peripheral urban locations as well as in the surrounding area.

CONNECTED, AUTOMATED AND PUBLIC TRANSPORT

Public transport has long played a key role within Vienna's mobility system, and the service has become even more vital, especially in recent years. In particular since the city's 2005 development plan (Dangschat/Hamedinger 2009: 104) and the subsequent 2025 concept, public transport has ultimately been seen as the backbone of the mobility system, and there is a belief that it needs to be bolstered further and made more attractive. By adopting the existing strategy plan for urban transport – the 2003 "Transport Master Plan" – Vienna

has decided to continue with the more offensive public transport policy that was launched in the 1990s. This stipulates that public transport should make up 40% of the modal split by 2020 (in 2001 it stood at 34%) while the share of MPT should be reduced to 25% by 2020 (2001: 36%; Stadt Wien 2006: 41). If we examine the most recent developments regarding modal split in Vienna (from 2003 to today), there is indeed a clear increase in public transport use as a percentage of modal split, its share rising by 10% to 39% (as of 2015). Like London, Budapest, Prague, Helsinki, Tallinn, Bucharest and Warsaw, Vienna has one of the highest levels of public transport use as a share of modal split across Europe's capitals (Nabielek et al. 2016: 26).

In order to see whether the addition of CAVs could improve existing public transport in peripheral urban and suburban areas (Gertz/Dörnemann 2016: 22), two automated shuttles have been undergoing trial runs in aspern Seestadt, a new urban development project in the northeast of Vienna, since the summer of 2019. The automated shuttles run along a two-kilometre route connecting areas in the south-west of Seestadt that are currently only moderately served by public transport with the terminus of one of the underground lines. The aim here is to find out just how much automated shuttles can be integrated into the service provided by the public transport authority, as well as to analyse their effectiveness as a link in the intermodal mobility chain. The project is also embedded within Seestadt's transport and road concept, which focuses on increasing the appeal of more environmentally friendly forms of mobility and striking a greater balance between them. Subsequent trials can be arranged to gain insight into the potential that may lie in supplementing public rail transport with flexible, needs-oriented, automated shuttles, especially within the city region.

CONNECTED AND AUTOMATED TRANSPORT IN THE LONG LEVEL 4

4

SETTLEMENT DEVELOPMENT, TRANSPORT POLICY AND PLANNING DURING THE TRANSITION PERIOD

© The Author(s) 2022
M. Mitteregger et al., *AVENUE21. Connected and Automated Driving: Prospects for Urban Europe*,
https://doi.org/10.1007/978-3-662-64140-8_4

4.1

TECHNOLOGICAL DEVELOPMENTS IN CONNECTED AND AUTOMATED VEHICLES: WHAT IS THE STATUS QUO?

Figure 4.1.1: **A LIDAR sensor on the roof of the test vehicle "Homer"**

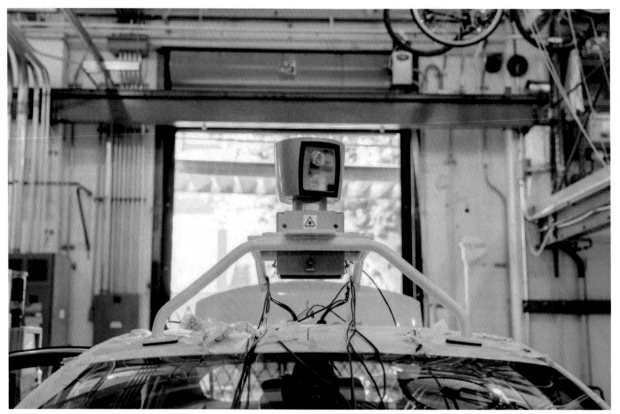

Source: voyage.auto

Connection and automation are the instrumental drivers of change in the transport system – yet they are two fundamentally different trends that are not necessarily related to one another (Perret et al. 2017: 6). Despite this, there is growing emphasis on their simultaneity and parallel development: whereas early research frequently spoke of autonomous driving or autonomous vehicles, more recent articles increasingly use the term "connected and automated vehicles": "Even though automated vehicles do not necessarily need to be connected and connected vehicles do not require automation, it is expected that in the medium term connectivity will be a major enabler for automated vehicles" (European Commission 2018: 4).

In response to the array of terms being used and their unclear meanings, SAE International (2018: 28) advises against the terms "self-driving vehicle", "autonomous vehicle" and "driving robot". What is commonly known as a "self-driving" or "autonomous" vehicle corresponds to the "fully automated vehicle" that is recommended in the SAE terminology and thus used here throughout.

The simultaneity of vehicles' connection and automation is explained by the fact that even at this stage increasing connection is seen as a prerequisite for some driving tasks. For instance, details about the current traffic situation, the condition of the roadway and possibly information from the infrastructure itself (traffic lights, tollbooths, etc.) may be necessary to facilitate safe automated operation of the vehicle (Ritz 2018: 184). Furthermore, some of the desired impacts of automated vehicles will only take hold when they act or cooperate in a connected way. Examples include the increased efficiency on roads and in the road network as a result of distribution, raising the vehicle flow rate and improving security (Kagermann 2017: 363; Shladover 2018: 196). That this view is held by the European Commission is evidenced by diverse initiatives and funding programmes (CAM, C-ITS, C-Roads; European Commission 2018: 4).

In addition, this means that although the majority of the first automated driving systems available, which only offer lower-level automated driving functions, are still relatively independent and not or hardly connected, it will be important in the long run – when higher levels of automation are achieved – for automated driving systems to be as connected as possible so that the desired effects are indeed achieved (Fig. 4.1.2; Shladover 2018: 193).

Figure 4.1.2: **Schematic diagram of the difference between ego-only, cooperative and automated systems**

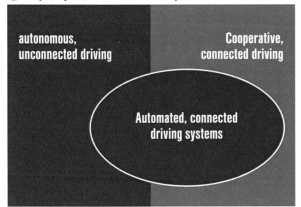

Source: AVENUE21 after Shladover (2018)

AUTOMATION LEVELS AND THEIR MEANINGS

Vehicles of different sizes are generally considered to be automated once they can perform a large proportion of dynamic driving tasks autonomously; the range of tasks undertaken culminates in fully driverless operation (Fig. 4.1.3). There are different automation level classifications for vehicles in passenger and freight transport, with the J3016 classification by SAE (Society of Automotive Engineers) having become established in the international scientific debate.

Those systems that the driver no longer has to supervise at all times because the longitudinal and lateral vehicle motion control is performed automatically in specific application cases (road types, speed zones and conditions) are known as conditional driving automation. However, the driver potentially has to be capable of taking over if the system requests them to intervene (Level 3). Highly automated vehicles, which are the subject of this research, are those whose systems can deal with

Figure 4.1.3: **Stages of automating driving systems**

	LEVEL	NAME	DESCRIPTION	VEHICLE MOTION CONTROL	AWARENESS OF SURROUNDINGS	FALLBACK LEVEL	OPERATIONAL DESIGN DOMAIN
			DRIVER PERFORMS ALL DYNAMIC DRIVING TASKS				
0		NO AUTOMATION	Driver drives independently, even if supporting systems are available.	DRIVER	DRIVER	NONE	N/A
1		ASSISTANCE SYSTEMS	Driver assistance systems help with operation of the vehicle during longitudinal or lateral vehicle motion control (not simultaneous).	DRIVER AND SYSTEM	DRIVER	DRIVER	LIMITED
2		PARTIAL AUTOMATION	One or more driver assistance systems help with operation of the vehicle during longitudinal and simultaneous lateral vehicle motion control. Driver must supervise the system at all times.	SYSTEM	DRIVER	DRIVER	LIMITED
			SYSTEM PERFORMS ALL DYNAMIC DRIVING TASKS				
3		CONDITIONAL AUTOMATION	Automated driving with the expectation that the driver must react to a request to intervene.	SYSTEM	SYSTEM	FALL-BACK-READY USER	LIMITED
4		HIGH AUTOMATION	Automated operation of the vehicle with the expectation that the driver will react to a request to intervene. Without human reaction, the vehicle continues to drive autonomously. The driver does not have to supervise the system at all times.	SYSTEM	SYSTEM	SYSTEM	LIMITED
5		FULL AUTOMATION	Fully automated driving, with the dynamic driving task being performed under all on-road and surrounding conditions as if by a human driver.	SYSTEM	SYSTEM	SYSTEM	UNLIMITED

AVENUE21 FOCUS

Source: SAE International (2018: 19)

Figure 4.1.4: **Operational design domain**

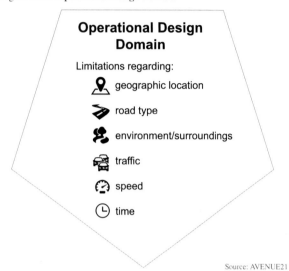

Operational Design Domain

Limitations regarding:

📍 geographic location

🛣 road type

🏃 environment/surroundings

🚗 traffic

⏱ speed

🕐 time

Source: AVENUE21

all situations automatically, but whose driverless operation only takes place in areas specifically designed for this purpose (operational design domain-specific); there is no request for driver intervention (Level 4). Finally, vehicles whose systems take over all driving tasks for all road types, speed zones and conditions (regardless of the ODD), and which are therefore operated from the start of the journey to its destination without a driver, are referred to as fully automated vehicles (Level 5; SAE International 2018: 19).

The operational design domain (ODD) describes the conditions under which an automated driving system functions. The parameters of these conditions may include geographic location, road type, surroundings, traffic, speed, time (see Fig. 4.1.4; SAE International 2018: 12).

LEVEL 4: AUTOMATED DRIVING SYSTEMS FOR SPECIFIC CONDITIONS (OPERATIONAL DESIGN DOMAINS)

The difference between automated driving systems (ADSs; SAE International 2018) that function under limited conditions (ODD-specific; Level 4) and those that function without limitations (ODD-unspecific; Level 5) truly is vast.

However, outside of these specific conditions or ODDs or when these conditions/ODDs change too dramatically, Level 4 automated driving systems are no longer functional (NHTSA 2017: 6). Should a transition to manual driving be necessary, then the driver is requested to take over the driving tasks. If they are not capable of doing so, the vehicle is returned to a risk-minimized system state (VDA 2015: 15; Wagner/Kabel 2018: 317).

Level 4 automated driving systems or applications have been in use since the 1990s. One well-known example in Europe is the park shuttle that connects the Rivium Business Park with the Rotterdam metro as a last-mile feeder line. The second generation of the park shuttle has been operating since 2006 over a distance of roughly 5 kilometres with the same number of stops. The limitations of the ODD were solved infrastructurally: the park shuttle drives on an asphalted route, which is separated from its surroundings on both sides by a one-metre-high fence and a hedge.

Yet in future, technological progress could see the operation of Level 4 automated driving systems without such major infrastructure measures as a structural divider or separate lanes (Hollestelle 2018: 24). Ultimately, such measures always serve to reduce the complexity of the automated driving system's ODD (e.g. avoiding interaction with cyclists) and hence the demands on the automated driving system.

SELECT LEVEL 4 AUTOMATED DRIVING SYSTEMS, THEIR OPERATIONAL DESIGN DOMAIN AND APPLICATIONS

Level 4 automated driving systems can have different use cases (Wachenfeld et al. 2015: 12), which are defined by the features of their respective ODD and other attributes, such as the possible application or use concept. Various Level 4 automated driving systems that are currently being discussed are designed for very different ODD and for different possible applications (Shladover 2018: 194). The result is numerous possible use cases for such automated driving systems (Fig. 4.1.5).

Figure 4.1.5: **Fields of application for Level 4**

MOTORWAY ASSISTANT

PARKING ASSISTANT

TOWN OR CITY ASSISTANT

AUTOMATED SHUTTLE BUS

Source: AVENUE21

A selection of current possibilities for Level 4 automated driving systems as well as their ODDs and applications are discussed below. The overview is limited to four applications.

1   MOTORWAY ASSISTANT

With the motorway assistant, the system exclusively takes over dynamic driving tasks on motorways or other trunk roads. During the journey on the motorway, no awareness of their surroundings is required of the passengers, who can instead occupy themselves with other activities (Wachenfeld et al. 2015: 12). The development of motorway assistants is primarily being encouraged by vehicle manufacturers, but is currently still in the development stage: for example, with the new A8, Audi is planning to market the first production vehicle with a motorway chauffeur (Level 3), which can only complete simple driving tasks in good weather conditions.

As initially described, the driver must therefore be capable of potentially taking over if the system requests them to intervene (Schrepfer et al. 2018: 32; Ritz 2018: 30). However, the further development of this system by vehicle manufacturers will lead to a constant increase in the possible dynamic driving tasks on motorways that the system can carry out to the point that a Level 4 motorway assistant is reached, which would allow the driver to read a book or sleep while the motorway assistant is driving on the motorway (Ritz 2018: 31). This would take the pressure off not only car drivers but also drivers of utility vehicles or lorries (Eckstein et al. 2018: 9) and be used for coaches travelling long distances.

Particularly with regard to lorries, motorway assistants are frequently discussed for lanes that are reserved for the exclusive use of lorries in order to reduce the demands on the driving system. These adaptations of the ODD take place with the aim of expediting the roadworthiness of such a Level 4 driving system (Shladover 2018: 194)

2   PARKING ASSISTANT (AUTOMATED VALET PARKING)

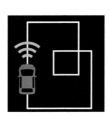

Evidently, vehicle manufacturers' focus in terms of automating passenger cars appears to be mainly targeted at motorways or motorway-like roads (Schrepfer et al. 2018: 34). However, manufacturers like Audi or Daimler are also woring to develop parking assistants/automated driving systems that enable automated valet parking (Ritz 2018: 30).

With automated valet parking, the automated driving system drives to a nearby or far-away parking space once the passengers have left the vehicle. The driver therefore saves time that would otherwise be spent searching for a parking space or actually parking the vehicle (Wachenfeld et al. 2015: 15; Shladover 2018: 194). At present, such automated valet parking is mostly being tested in large car parks, i.e. not in the road network but in places that can be described as distinct settings: Daimler and Bosch, for example, are testing such a system in an indoor car park in Stuttgart, where the vehicle is simply left in a drop-off area and then parks automatically after a command has been activated on the driver's smartphone and automatically drives to a pick-up area after another smartphone command has been selected (Daimler 2018).

In future, such automated driving systems might operate not only in distinct settings, but also in defined and approved areas of the low-capacity road network (e.g. inner-city areas). That would allow the driver to stop right outside a restaurant, for example, and then task the vehicle with automatically finding, driving to and parking in a free parking space once they have got out of the vehicle (Eckstein et al. 2018: 9). However, this would mean that the parking assistant would closely resemble a city assistant (see next paragraph). Due to the advantages associated with this automated driving system (the vehicle picks up its passengers nearby), the development of the parking assistant is being considered not only for private cars, but also – on the part of the manufacturers (e.g. BMW and Daimler) – increasingly in connection with considerations of their own car-sharing services (Drive-Now and Car2go; Ritz 2018: 114; Lenz/Fraedrich 2015: 185).

3   TOWN OR CITY ASSISTANT

Vehicle manufacturers, which are increasingly presenting themselves as mobility providers, are also working – partly together with new players on the mobility market like Uber or Waymo – on automated driving systems for quite "urban" conditions or defined parts of the lower-capacity road network (Ritz 2018: 135). Current examples include the development of a city assistant by Audi, the tests by Waymo in a designated 100 square-mile area in Chandler, a (suburban) outlying district of Phoenix in Arizona, and the announcement by Daimler and Bosch that they will jointly release Level 4 vehicles in an urban setting over the coming decade (Hawkins 2017; Ritz 2018: 30; Daimler 2018).

In this case, the automated driving system – often referred to as a town or city assistant – takes over the driving tasks in a defined and approved part of the lower-capacity road network. The driver thus becomes a passenger in this setting (Wachenfeld et al. 2015: 17; Altenburg et al. 2018: 4).

As a result of the advantages of automated driving systems in this context (e.g. the possibility for a door-to-door service; Lenz/Fraedrich 2015: 185), such systems are increasingly related to concepts like car sharing and ride sharing and then frequently called automated private or shared taxis. The boundaries between MPT and PT are more and more fluid, because in such use cases both vehicle manufacturers and new players on the mobility market see the opportunity for new business models, while public transport companies could expand their service provision by offering such flexible automated driving systems, especially in the suburbs. In principle, the aim with such automated driving systems is that the passenger can contact the control centre in an emergency and pay for the journey via smartphone (Eckstein et al. 2018: 9).

Such automated driving systems are not only being discussed for passenger but also for goods transport, especially as a solution for the last mile. They are intended to take over the last mile from an inner-city warehouse or offline shop to a customer in a defined and approved part of the lower-capacity road network. One example are the vehicles frequently referred to as delivery robots by the manufacturer Starship Technologies, which have a container in which a parcel can be placed (Vogler et al. 2018: 152). Currently, the majority of such automated driving systems are still in operation in business parks (e.g. in Mountain View, USA) or other special zones (where there may have been such automated driving systems, such as for in-house transport, for a long time; Flämig 2015: 378; Hern 2018). Here, too, it is a matter of reducing the demands made of the automated driv-

ing system. Generally, however, such driving systems are intended to operate at a speed of max. 6 km/h on footpaths in defined, approved areas. They have already been tested in this form in Hamburg (together with Hermes) and Düsseldorf, among other places.

4 AUTOMATED SHUTTLE BUS

At present, automated shuttle buses are primarily being tested by public transport companies. Examples include the shuttle bus tests by the Deutsche Bahn (DB) in Bad Birnbach, by the Wiener Linien in Vienna and by the PostAuto Schweiz AG in Sitten in Switzerland. The best-known manufacturers of these automated shuttle buses are the companies Navya and EasyMile. The shuttle buses are mostly designed for a capacity of 8 to 12 passengers and according to their manufacturers can reach a maximum speed of 45 km/h (Navya 2017: 13); nevertheless, most of them are operating at speeds of just 15 to 20 km/h (Zankl/Rehrl 2017: 38; Postauto Schweiz AG 2016).

The automated shuttles are predominantly being tested by the public transport companies as possible feeders for the main underground and suburban railway lines and hence as an expansion of the PT network, for example in the suburbs (Michelmann et al. 2017: 2). The potential personnel-related cost savings as well as the more flexible deployment of such vehicles due to their smaller sizes are viewed as financial relief for the operation of PT in such areas (Lenz/Fraedrich 2015: 191;

Figure 4.1.6: **Select automated driving systems and their operational design domain**

AUTOMATED DRIVING SYSTEM (ADS)	OPERATIONAL DESIGN DOMAIN		
	MAIN SETTING	**SPEED**	**OTHER ROAD USERS**
MOTORWAY ASSISTANT	Regulated-access motorways or other trunk roads (at first presumably only with light traffic volumes)	Due to the vehicle type, the maximum speed must be 60 km/h or above.	In mixed traffic with other motorized road users (no non-motorized road users)
PARKING ASSISTANT	Distinct settings like indoor and outdoor car parks (subsequently defined and approved parts of the lower-capacity road network)	Low speeds	In mixed traffic with other – also non-motorized – road users
TOWN OR CITY ASSISTANT	Defined and approved parts or areas of the lower-capacity road network	Speeds according to the defined and approved parts or areas of the lower-capacity road network	In mixed traffic with other – also non-motorized – road users
AUTOMATED SHUTTLE BUS	Predefined routes or lines (with clearly visible markings) in the lower-capacity road network	Low speeds (max. 20–30 km/h)	Possibly separate from other road users

Haider/Klementschitz 2017: 7; Eckstein et al. 2018: 9). It can be assumed that automated shuttles will only run on select, defined, approved and clearly marked routes (and lanes) at first and stop at predefined stations. In these areas, they can operate without drivers, but will be supervised and where necessary manoeuvred by traffic controllers in a control room, similar to the established procedure in aviation (Eckstein et al. 2018: 8). In the course of continuous technical development, the area of operation for such automated driving systems might ultimately extend to defined and approved parts of the lower-capacity road network. The automated driving system of automated shuttle buses would thereby approximate a city assistant – combined with the concept of ride sharing.

SUMMARY

In summary, the range of automated driving systems described above can be categorized on the basis of the geography or setting in which they can operate, which comprises several parameters of the ODD. In simple terms, these categories are: (1) motorways and other trunk roads, (2) parts of the lower-capaci-

ty road network, (3) routes in the lower-capacity road network and (4) distinct settings (Fig. 4.1.6). Whereas the motorway assistant focuses exclusively on motorways and other trunk roads, the parking assistant (valet parking) functions in the first instance only in distinct areas like car parks. In contrast, the so-called city assistant concentrates on defined and approved parts of the lower-capacity road network, while the automated shuttle bus runs primarily on defined routes in the lower-capacity road network. In future, Level 4 automated driving systems might also cover combinations of these scenarios. Technical developments will ultimately culminate in a Level 5 automated driving system, which will function in all contexts or settings (ODD-unspecific), and (to the greatest possible extent) without limitations (Shladover 2018: 195).

The Level 4 automated driving systems described here are relevant for both passenger and freight transport, but only in distinct sections of the road network (Fig. 4.1.7).

The motorway assistant can be used for both private cars and logistics vehicles like lorries and vans, and even coaches. The parking assistant is primarily being discussed in the context of private cars or A-segment

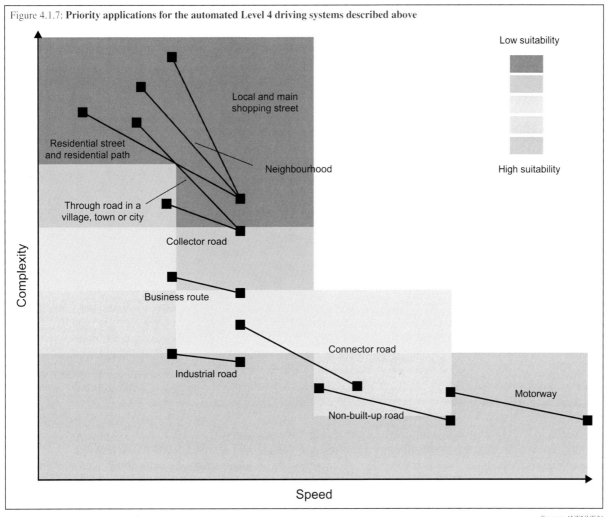

Figure 4.1.7: **Priority applications for the automated Level 4 driving systems described above**

Source: AVENUE21

vehicles like pods or LSEV (low-speed electric vehicles), as well as in connection with car sharing (private taxis). The so-called city assistant can be used both for private cars and A-segment vehicles, but also for car sharing (private taxis) and ride sharing (shared taxis) – the boundaries are blurred here between individual and collective passenger transport – as well as for (last-mile) logistics vehicles. The automated shuttle bus is being considered in public transport, where it would have pre-established or flexible stops and operate as a free-floating or route-based system.

THE IMPORTANCE OF LONG LEVEL 4 FOR URBAN AND MOBILITY PLANNING

The term ODD was only adopted by SAE in 2016 and has since been growing in significance: both in the development of the technology and in urban and mobility planning. A prolonged transition period in which automated vehicles are possible and in use, though only in specific environmental conditions, results in a staggered time frame. This staggering is a consequence of the complexity that characterizes urban streetscapes in particular. The opinion that CAVs will be implemented homogeneously in cities is hence outdated. Rather, various applications will be implemented in selective spaces in European cities and will hence only be available to certain sections of the population. Furthermore, this will result in a possible shift in location quality, with some locations being connected to a supra-regional transport network and others not. Consequently, the possible changes to the city as a result of automated vehicles will only take place in individual areas at first (Ritz 2018: 74). Chapter 4.4 takes a close look at this issue. As shown in the scenarios in Chapter 5, this limitation raises both the need and scope for action by urban and mobility planners.

CONNECTING VEHICLES

In the context of vehicles' automation, their connection with one another or with their surroundings is increasingly playing a major part (Rammler 2016: 14; Bönninger et al. 2018: 97). In the majority of the applications described above, especially car and ride sharing, connection is essentially the prerequisite for users being able to access the automated vehicle via app or web portal (Johannig/ Mildner 2015: 4).

Vehicles' capacity to connect with one another is categorized according to the thing or person with which/whom the connection is established. "Vehicle to everything" or V2X (the sum of all functions in Fig. 4.1.8) describes vehicles' complete communication ability: whether with one another (V2V), with infrastructure (V2I) or with pedestrians' mobile devices (V2P; Shladover 2018: 191).

The possible applications range from dynamic and highly concentrated (goods) platoons as well as connected collision warnings and danger alarms (V2V) to real-time information about weather and road conditions (V2I) to shuttles that communicate their status directly with pedestrians or predictively with pedestrians' mobile devices (V2P; Owens et. al 2018: 71; Shladover 2018: 191). Moreover, users should have the option to be seamlessly mobile from departure to arrival depending on their situation (Boban et. al 2017: 2). Figure 4.1.8 provides an overview of the different connection types and their possible applications.

The prerequisite for this is the reliable and stable, and most importantly highly efficient and quick, sharing of information or data on the basis of communication technologies, sensors and network connections (Maracke 2017: 64). The industry differentiates here between long and short latency times. While the latter primarily concern collision warnings, speed limits or electronic payments for parking and tolls, the former are mainly related to infotainment and traffic information services on long journeys.

Various wireless communication technologies can be used to share this data (Shladover 2018: 192). The most applicable technologies are ITS-G5 (WLAN IEEE 802.11p), cellular networks (LTE-Vehicular/LTE Advanced or in future 5G) and digital broadcasting such as DAB (Digital Audio Broadcasting), DAB+, DMB (Digital Multimedia Broadcasting) or DAB-IP. At present, it is still entirely uncertain which communication technology will prevail in the context of automated driving systems: while the European Commission is pursuing the idea of a complementary communication mix with the application of hybrid communication technologies, in the USA a single (short-range) communication technology is preferred by the National Highway Traffic Safety Administration in draft legislation on the ITS-G5 standard (Sänn et al. 2017: 62).

As a result of the growing importance of vehicles' connection, the requirements for security and data protection are being raised (Lemmer 2015: 61). Every connected vehicle collects considerable amounts of at times sensitive data and information on movement patterns, personal travel habits or financial matters, which must be not only saved but also analysed and secured. Without security standards, the more automated and connected vehicle systems become, the more vulnerable they will be to outside attacks and to malfunctions (Seider/Schmitz 2017).

Regarding communication from vehicle to vehicle or between the vehicle and its manufacturer's servers, it must therefore be ensured that there is adequate protection from attacks by hackers, that data integrity is guaranteed and that communication is robust. In addition, it is imperative that the systems cannot simply be stopped by denial-of-service attacks (e.g. a distributed denial of service or DDoS; Ritz 2018: 205). Ultimately, the topic of connection – especially in light of user acceptance – is closely related to guaranteeing data privacy and protection from cyberattacks (Seider/Schmitz 2017 and Chap. 3.4).

THE IMPORTANCE OF CONNECTION FOR URBAN AND MOBILITY PLANNING

The issue of connection is significant for cities. Connection with infrastructure (traffic lights, traffic information or control systems) means a considerable financial outlay (Mitteregger et al. 2019). To organize and access multimodal mobility services, the connection of vehicles as well as different operators' fleets is crucial. If cities want to play a part in organizing sharing services, then access to data is fundamental. Data will also be of growing importance in future for traffic information systems and to steer flowing traffic by means of toll systems. These data can in turn be generated by municipal investments in digital infrastructure or by requiring the operators of sharing fleets to communicate this information with the city (Chap. 3.3).

Figure 4.1.8: **Overview of connection types and their possible applications**

CONNECTION TYPE	EXPLANATION	APPLICATIONS
 **Vehicle to Vehicle (V2V)**	Communication between vehicles	Real-time information about: • collision warning • danger alarms • cooperative adaptive cruise control (CACC) • platooning • connection guarantee when changing transport modes • etc.
 **Vehicle to Infrastructure (V2I)**	Communication with roadside infrastructure	Real-time information about: • weather • road conditions • traffic lights (e.g. green wave) • variable speed limits • toll payments • etc.
 **Vehicle to Pedestrian (V2P)**	Communication with passengers and non-motorized road users	Real-time information about: • position • speed • direction • etc.

Source: AVENUE21 after Shladover (2018: 191) and Perret et al. (2017: 16)

4.2

SETTLEMENT AND INFRASTRUCTURE ASPECTS OF SPATIALLY SELECTIVE IMPLEMENTATION

The development of settlements and cities is closely connected to transport or rather to technological innovations in mobility. While journeys were almost exclusively undertaken on foot until the mid-19th century, only a short time later horse-drawn carriages and buses, trains, trams, undergrounds and then cars became the main modes of transport and accordingly left their mark on settlement development (Fig. 4.2.1; Safdie/Kohn 1998: xii). Today, it is easy to draw a connection between mobility technologies and the development of European cities, their variety of historic, medieval districts and more recent neighbourhoods. Consequently, the evolution of settlement structures – meaning settlements' expansion, inner structure and spatial distribution – reflects the historical development of transport systems: the means of transport available, the popular transport routes and above all the speeds of transport modes (Schmitz 2001: 27).

Both the vehicles themselves and the transport infrastructure (rail and road networks, harbours and airports) were the decisive factors regarding which mobility innovations would have an influence on life in and the structure of settlements. The history of infrastructure is also one of innovations, which frequently go unmentioned (McShane 1994). Infrastructure is the immobile component that is essential if new modes of transport are to have a spatial impact (changes to accessibility and land use as well as associated economic and social

effects). (In Europe) it is often the public sector that makes funding available for the construction and maintenance of transport infrastructure. The economic sense of investments in transport infrastructure is therefore a much-discussed transport policy issue (Aschauer 1989, Deng 2013). This aspect will be the focus of our study of CAT in SAE Level 4 in this section. As Level 4 driving systems can only be implemented in parts of the road network, they may require investments in or expansion of infrastructure.

4.2.1 SETTLEMENT DEVELOPMENT AND MOBILITY INNOVATIONS: LOOKING BACK

In the Middle Ages, European cities were mostly characterized by high density, narrow streets and crowdedness. This resulted from the necessity for fortifications around the city, which in turn were needed because of the relative spatial proximity of cities to one another, and from the fact that most journeys had to be undertaken on foot (Mumford 1984; Wegener/Fürst 1999: 4). The pedestrian network remained the transport system that defined urban development until well into the 19th century. These cities' structure accordingly revolved around walking distances: cities were small and cohesive urban units and the diameter of the urban area

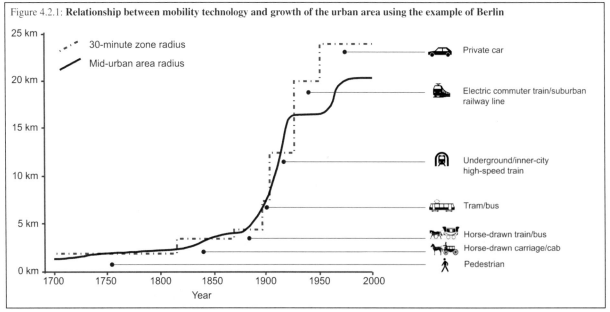

Figure 4.2.1: **Relationship between mobility technology and growth of the urban area using the example of Berlin**

- - - 30-minute zone radius
—— Mid-urban area radius

Private car

Electric commuter train/suburban railway line

Underground/inner-city high-speed train

Tram/bus

Horse-drawn train/bus
Horse-drawn carriage/cab
Pedestrian

Year

25 km / 20 km / 15 km / 10 km / 5 km / 0 km

1700 1750 1800 1850 1900 1950 2000

Source: AVENUE21 after Kagermeier (1997: 25) and Lehner (1964: 22–23)

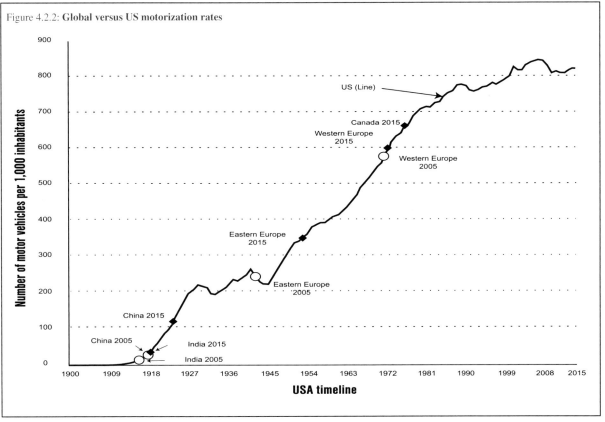

Figure 4.2.2: Global versus US motorization rates

Source: Davis et al. (2015: 3–8)

rarely exceeded five kilometres (Kainrath 1997: 16). Such structures can still be found today in the medieval centres of numerous European cities (Newman/Kenworthy 1999: 28). With the advent of horse-drawn buses, trams, supra-regional trains, wooden and iron rails and new road surfaces (stone, brick and wooden paving, tarmac and finally asphalt) in the 18th century, accessibility and subsequently settlement structure changed (Kainrath 1997: 16). A fundamental shift in the social significance of mobility and the streetscape took place (McShane 1979: 57–80).

The last far-reaching change in mobility was brought about by the growing popularity of the private motor car over the course of the 20th century. Beginning in the USA (where asphalt road surfaces also spread more rapidly), it spread to Europe after World War II. The extensive accessibility afforded by cars now made it possible to use areas between railway lines for city expansion (Wegener/Fürst 1999: 5). In addition, the general increase in prosperity associated with the postwar economic growth in Europe led to widespread individual motorization. The prevalence of private cars simultaneously enabled and encouraged the spatial separation of functions like living and working (Kagermeier 1997: 24). The result was less organized and more dispersed urban growth accompanied by high levels of urban sprawl on the outskirts (Wegener/Fürst 1999: 5). The urban fringe is still dominated by car travel today as there are considerable gaps in public transport provision (Kainrath 1997: 16).

4.2.2 RATE OF CHANGE: THE DIFFUSION OF TRANSPORT TECHNOLOGIES

Before transport technologies can have a spatial impact, they must first be accepted by a growing number of people and be used in everyday life. These are long-term processes – a fact that is often overlooked (King/Baatartogtokh 2015) by the theory of disruptive technologies (Christensen 2003). The diffusion of innovations (Rogers 2003) is a communicative process during which individuals or social groups choose to adapt and hence to accept the effort that goes along with such change. The diffusion starts slowly, gathers pace and then loses speed as soon as saturation effects emerge. This has proven to be a valid finding (Kucharavy/De Guio 2011, Grubler et al. 2016).

The duration of technological diffusion processes depends on a range of factors. They include the complexity of the technology, the length of the "formative phase" (Bento/Wilson 2016), the effort needed to change existing habits, the amount of investment required for its introduction, the expected advantages for various groups and whether legal amendments or new regulations are necessary.

Figure 4.2.3 summarizes the characteristics of long diffusion processes as described by Grubler et al. (2016) and applies them to CAT. The high global prevalence of road traffic and the large number of individuals who will be affected by changes to transportation and the streetscape are principal factors affecting the duration of the diffusion process for CAT. In addition, there is considerable need for coordination between various political levels and stakeholders (see Chap. 4.6), e.g. in order to certify diverse automated driving functions and vehicle types (Walker 2016). Moreover, standards for the physical and digital infrastructure and to ensure certain ODDs must be coordinated and implemented (European Commission 2017, 2019). Furthermore, major investments are expected to be necessary at various levels of the road network, which will present a not insignificant challenge to states and municipalities (POLIS 2018, Mitteregger et al. 2019). In terms of the technological development, the amalgamation of different sensor data to produce a coherent picture is a key challenge, which has not been required to this extent for any other use case (see Chap. 4.4).

ISSUES OF AUTOMATIC DRIVING SYSTEMS' ACCEPTANCE

The response to the question of CAVs' acceptance is determined by the duration of the diffusion process. This affects not only the travellers (passengers), but also all other road users. It should not be underestimated that with Level 4 "driving robots", the first time people have to interact with automated mobile machines in public space is in critical (and potentially lethal) situations. To date, this experience is still limited to those working in sectors like logistics (e.g. in ports or logistics centres), agriculture, mining or the military (drones). Figure 4.2.4 summarizes the principal areas of acceptance currently being discussed in the specialist literature.

CAV passengers must learn to accept that they are being driven by a software-controlled machine that steers and supervises all activities during the journey, and that can potentially be affected or directed by external entities (police, mobility service provider, infrastructure operators, but also hackers). Interacting with the vehicle as a passenger will be a major hurdle for certain social groups (trust, familiarity with technology, the digital divide). Yet also for other road users who will have to adjust their behaviour in public space to that of the automated driving system, there is a pressing need to establish trust (which is being addressed in current test situations by means of signals on the vehicles). In the case of CA ride sharing, the relatively small space inside the vehicle is shared with strangers for the duration of the journey without the presence of a driver, a factor that – in light of increasing social differentiation – will lead to acceptance being low and potentially sinking in sparsely populated areas and in off-peak periods at first (Merat et al. 2017). This applies particularly to groups with limited mobility who are supposed to benefit the most from these shuttle services.

A low acceptance of automated vehicles might lead to behaviour in the public space of the street once again changing fundamentally (as was the case with the advent of cars, see Chap. 3.2). During operation, automated driving systems will gather comprehensive data in public space – including on people's behaviour in the streetscape – and utilize them in the context of data-based business models or traffic management. Comprehensive surveillance is rarely accepted in European (in contrast

Figure 4.2.3: **Comparison of the characteristics of long technology diffusions and connected and automated driving systems**

LONG TECH. DIFFUSIONS	CONNECTED AND AUTOMATED DRIVING SYSTEMS
Adaptations in organizational and institutional settings, of several technologies and of infrastructure become necessary.	• Institutions and organizations first have to create the legislative framework for CAVs and develop criteria for their authorization or certification (Schoitsch et al. 2016). As a result of transnational transport networks and the standardized production of vehicles, there is a fundamental need for coordination at every political level.
	• Connected and automated vehicles replace non- or minimally automated vehicles in pre-existing infrastructure. Even if part of the existing infrastructure can be used, it nevertheless requires major adaptations.
New technological and social concepts have to be developed or learned.	• The main technical challenge of CAT is amalgamating different sensor data to produce a coherent picture. This picture has to be interpreted by machines (via artificial intelligence) and serves as the basis for traffic control and driving decisions (see Chap. 4.6).
	• The interaction of non-human actors with other road users poses new socio-psychological challenges, which must be addressed and accepted in the long term (Merat et al. 2017, Rogers 2003).
Significant investments in widespread technologies and infrastructure become necessary, with the expense of the adaptation only paying off at a later stage.	• Investments could become necessary at every level of the transport network. Of relevance are the sizes of the motorway and trunk road networks (1.9% of the transport network in Austria), dual carriageways (29.3%) and municipal roads (68.73%; BMVIT 2018) and those responsible in each case for their maintenance and repair, as well as different kinds of funding (e.g. via toll or transfer payments).
	• In addition, further investments may be necessary in the digital infrastructure, the connection of the vehicles (DG MOVE 2016: 41). The scale of these investments differs greatly from place to place and is not currently foreseeable.

Source: AVENUE21 after Grubler et al. (2016: 19)

to Chinese) cities at present. Closely connected to the introduction of CAVs is the transformation from passive to active security systems, which will fundamentally alter the understanding of public streets in European cities in future (Mitteregger 2019).

4.2.3 THE IMPORTANCE OF TRANSPORT INFRA-STRUCTURE DURING THE LONG LEVEL 4

Transport infrastructure, and road infrastructure in particular, is deemed to be vitally important for the economic development of regions (Aschauer 1989). Consequently, the planning and construction of transport infrastructure is considered a key structural policy measure. The European Union has allocated some €600 billion to "complete and modernise a true trans-European network" by 2020 (European Commission 2005: 3). From a structural policy perspective, the European initiative "Cooperative Intelligent Transport Systems" (C-ITS) is believed to be one of the key measures to get CAM off the ground. It covers a broad spectrum of infrastructure applications (from traffic control and management systems to blanket 5G coverage). C-ITS are intended to increase road safety, boost efficiency and improve comfort (European Commission 2016: 3). Especially in Europe and regarding CAT, the widespread perception among policymakers and planners that roads lead to wealth (Deng 2014: 687) should be viewed in a more nuanced light.

SETTLEMENT DEVELOPMENT THROUGH TRANSPORT INFRASTRUCTURE

The relationship between the effects of transport infrastructure on land use and settlement structure has been confirmed by recent overview studies focusing on various spatial criteria (Deng 2014, Kasraian et al. 2016). However, this relationship is generally more nuanced than usually assumed (Fig. 4.2.5). There is a demonstrable correlation between the extent of development in the settlement area, its current accessibility and the effectiveness of transport infrastructure investments as a structural policy measure. Saturation effects become clear, which arise in areas that are easily accessible and/or already have a well-developed settlement structure. Consequently, the strongest impacts are to be expected primarily in areas which are not fully developed and to which access can be improved by the construction of new transport infrastructure.

Another feature of improved accessibility as a result of transport infrastructure measures is that, from an economic perspective, it does not benefit all sectors equally (Deng 2014: 691–692). Transport-heavy sectors like logistics and construction profit much more than e.g. the textile industry (Fernald 1999: 628). Cantos et al. (2005) were also able to demonstrate an unequal distribution of accessibility effects with regard to companies' sales markets. For example, those branches that market their products at national or transnational level (indus-

Figure 4.2.4: **Criteria discussed in the literature regarding passengers' and other road users' acceptance of SAE Level 4**

CRITERION	STUDIES
PASSENGER	
"Being transported"	Hancock et al. 2011, Malodia/Singla 2016, Schaefer/Straub 2016
Ride sharing	Ahmadpour et al. 2016, Beirao/Sarsfield-Cabral 2007, Chan/Shaheen 2012, Dueker et al. 1977, Malodia/Singla 2016, Merat et al. 2017, Thompson et al. 1991, Venkatesh et al. 2003
Surveillance during the journey	Crittenden 2017, Litman 2017, Schulz/Gilbert 1996
Human-machine interaction	Grush et al. 2016, Hoff/Bashir 2015, Merat et al. 2017, Schaefer/Straub 2016, Seppelt/Lee 2007, Venkatesh et al. 2003, Wiseman 2017
Access to external entities	Anderson et al. 2016, Gontar et al. 2017
OTHER ROAD USERS	
Empty runs	Elliot/Long 2016
Human-machine interaction	Anderson et al. 2016, Grush et al. 2016, Hoff/Bashir 2015, Merat et al. 2017, Parkin et al. 2016, Rodriguez et al. 2016, Schaefer/Straub 2016, Seppelt/Lee 2007, Venkatesh et al. 2003
Surveillance on public streets	Anderson et al. 2016, Cirittenden 2017, Schulz/Gilbert 1996

try, manufacturing trades) benefit, while sectors operating regionally (retail, construction, the service industry, agriculture) tend to face losses in profits as a result of improved accessibility.

4.2.4 CAT AT THE INTERSECTION OF LAND USE AND TRANSPORT

There is an interdependency between land use and transport and via the key parameter of accessibility, together they form a complex of interrelated effects (Fig. 4.2.6; Wegener/Fürst 1999: 5–6; Bertolini 2012: 19). The effects within the system happen at different speeds: if there is a change in the transport system, this causes an immediate change in accessibility and simultaneously causes corresponding changes in the perception of distances (Kagermeier 1997: 22). As a result, adjusted mobility charges can give rise to new or different functional relationships between pre-existing locations. Similarly, a change in land use or settlement structure leads to a change in accessibility and to a relatively fast shift in mobility activities (within a matter of years or even days; Bertolini 2012: 2).

However, the respective changes in accessibility only have a medium- to long-term impact (over decades) on settlement structure or transport services (Bertolini 2017: 27). Consequently, the change in effort needed to cover distances only has a medium- to long-term effect on construction investors' location decisions (and on their construction work) and on individuals, households and companies, and hence on settlement structure (Kagermeier 1997: 22).

Figure 4.2.5: **Spatial impacts resulting from transport infrastructure networks and land use**

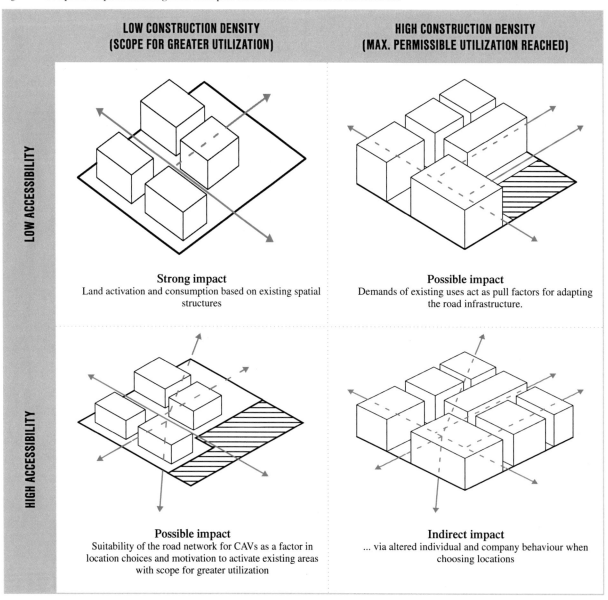

LOW CONSTRUCTION DENSITY (SCOPE FOR GREATER UTILIZATION)

HIGH CONSTRUCTION DENSITY (MAX. PERMISSIBLE UTILIZATION REACHED)

LOW ACCESSIBILITY

HIGH ACCESSIBILITY

Strong impact
Land activation and consumption based on existing spatial structures

Possible impact
Demands of existing uses act as pull factors for adapting the road infrastructure.

Possible impact
Suitability of the road network for CAVs as a factor in location choices and motivation to activate existing areas with scope for greater utilization

Indirect impact
... via altered individual and company behaviour when choosing locations

Source: AVENUE21 after Kasraian et al. (2016)

As a mobility innovation, connected and automated vehicles have enormous potential to transform the transport system. They make different transport services possible, which will lead to changes in the demand for transport (Fig. 4.2.2; Alessandrini et al. 2015: 148; Friedrich/Hartl 2016: 7). In the long term, impacts on urban and settlement development are likely (European Commission 2016: 2). Based on the assumption of a Long Level 4 in which CAVs can only be used in parts of the transport network, a highly dynamic transport and spatial system can be expected. The subsequent consideration of global simulation studies should be read with this in mind.

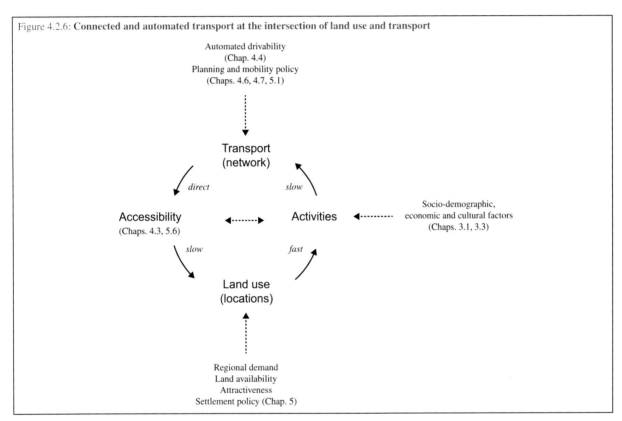

Figure 4.2.6: **Connected and automated transport at the intersection of land use and transport**

Automated drivability
(Chap. 4.4)
Planning and mobility policy
(Chaps. 4.6, 4.7, 5.1)

Transport
(network)

direct   *slow*

Accessibility   Activities   Socio-demographic,
(Chaps. 4.3, 5.6)   economic and cultural factors
                                                                              (Chaps. 3.1, 3.3)

slow   *fast*

Land use
(locations)

Regional demand
Land availability
Attractiveness
Settlement policy (Chap. 5)

Source: AVENUE21 after Wegener/Fürst (1999) and Bertolini (2012)

4.3

RESEARCH ON FULLY AUTOMATED VEHICLES' IMPACTS ON THE CITY: STATUS QUO

Figure 4.3.1: **Collated results of the simulations from recent studies**

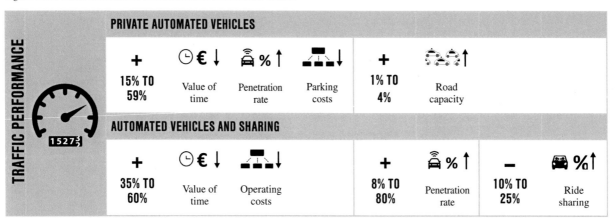

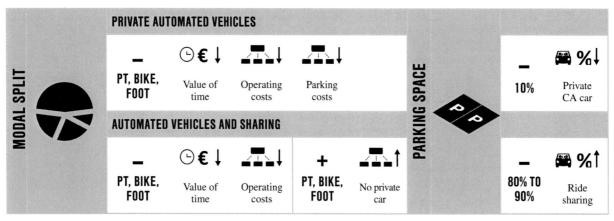

Source: AVENUE21

At most, higher-level automated vehicles (simulation studies do not analyse vehicles' connection, which is why this section only discusses automated vehicles) are on the road exclusively in the context of test projects at present. In order to investigate and assess the possible impact of automated vehicles on cities in the future, numerous studies therefore rely on simulations: by using various assumptions and scenarios regarding the forms of transport – including automated vehicles – available in future, automated vehicles' possible impact on traffic and urban spaces is simulated with the aid of computers. In addition, there are studies that attempt to assess the effect of automated vehicles on issues such as social justice/inclusion or municipal budgets.

Figures 4.3.1 and 4.3.6 provide an overview of the results of recent studies on the aforementioned issues and a short explanation of the accompanying effects.

Among other things, they comprise road capacity/traffic jams, road safety, transport infrastructure, vehicle ownership, transport demand, parking spaces, settlement structure, social justice, health, the environment, cyber security, the economy and governance (Milakis et al. 2017: 6).

For the purposes of this project, an analysis was conducted of the effects of automated vehicles on transport demand and settlement structure or land use. In total, 37 modelling studies were analysed from various countries around the world, though primarily the USA and Europe (Soteropoulos et al. 2018a). The results show that the way automated vehicles are used as well as the associated modelling assumptions (e.g. proportion of car and ride sharing, value of time, increase in road capacity) are particularly important.

Figure 4.3.2: **Analysed impact of connected and automated transport on transport and urban spaces in simulation studies**

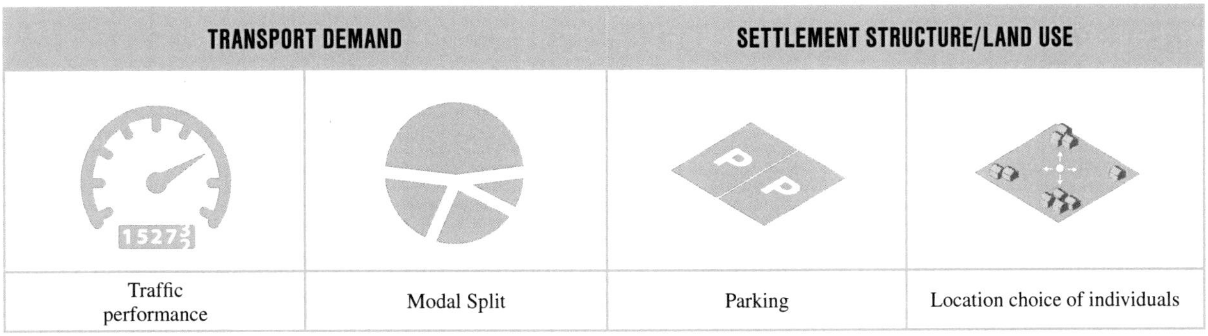

TRANSPORT DEMAND		SETTLEMENT STRUCTURE/LAND USE	
Traffic performance	Modal Split	Parking	Location choice of individuals

Source: AVENUE21

4.3.1   IMPACT ON TRANSPORT AND URBAN SPACES

Present simulations show that there are mainly four aspects that can be differentiated with regard to the impact of CAVs on transport and urban spaces, with the subject of investigations being the demand for transport on the one hand and the influence on settlement structure or land use on the other (Fig. 4.3.2). Further, these studies consider effects on traffic performance, the modal split and parking spaces (Fig. 4.3.1) as well as individuals' and companies' choice of location (Fig. 4.3.3).

TRAFFIC PERFORMANCE

Regarding traffic performance, it is clear that private automated vehicles are overwhelmingly accompanied by a rise in the number of kilometres driven due to shifts from other modes of transport, with traffic performance growing by 15% to 59% when it is assumed that there is a substantial reduction in the value of time (which includes perception of time) and in parking costs and that automated vehicles have a high market share. Even automated vehicles associated with sharing overwhelmingly lead to an increase in the kilometres driven due to shifts from other modes of transport and empty runs. This increase lies between 35% and 60% when it is assumed that there is a reduction in the value of time and lower costs for their use, and between 8% and 89% when it is assumed that part or all MPT demand is covered by automated vehicles combined with sharing services. Assuming a high proportion of ride sharing results in a 10% to 25% reduction in traffic performance.

MODAL SPLIT

In terms of the modal split, it is evident that private automated vehicles will overwhelmingly lead to a reduction in the proportion of PT, bikes and pedestrians in the modal split (see also the expert opinions and assessments in Chap. 3.4), with substantial reductions being especially clear when a sharp decrease in the value of time and in parking and operating costs is assumed. Even automated vehicles associated with sharing services overwhelmingly lead to a drop in the share of PT

passengers, cyclists and pedestrians in the modal split, especially when marked reductions in the value of time and low use costs are assumed. Increases in the proportion of PT use, cycling and walking in the modal split only arise when it is assumed that use costs will be quite high and there will be no private vehicles.

PARKING SPACES

With regard to a possible reduced need for parking spaces, it is clear that automated vehicles combined with a high proportion of sharing, especially ride sharing, could decrease the overall number of vehicles by approximately 90%, which could also lead to a reduction by as much as 80% to 90% in the need for parking spaces. In contrast, private automated vehicles, where sharing only takes place between the members of a single household, might only cut back parking spaces by about 10%. This would also result in a significantly lower reduction in the need for parking spaces and hence fewer potentially usable areas being made available.

LOCATION CHOICE OF INDIVIDUALS AND COMPANIES

Concerning a change in individuals' and companies' choice of location (Fig. 4.3.3), it becomes apparent that – especially when a reduction in the value of time in the vehicle and an increase in capacity are assumed – private automated vehicles lead to population growth in well-connected suburban/rural areas, i.e. tend to cause sprawling urban growth. In contrast, efficient PT as a result of automation (e.g. connected and automated shuttles for the last mile) leads to population increase in urban areas and hence tends to encourage urbanization processes.

Automated vehicles with sharing services might also curb urban sprawl and suburbanization processes, although it does appear that some demographic groups might move further away from the city centre as a result of the improved transport access. Furthermore, automated vehicles with sharing services could aggravate the deindustrialization trend in cities, i.e. contribute to yet more secondary-sector companies shifting their locations to areas outside of the city.

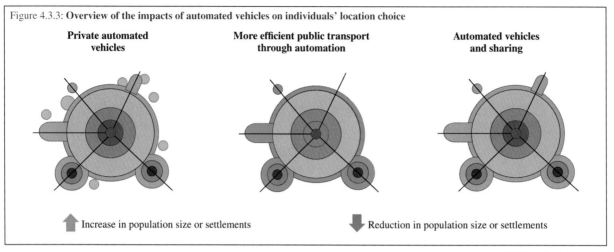

Figure 4.3.3: **Overview of the impacts of automated vehicles on individuals' location choice**

Private automated vehicles

More efficient public transport through automation

Automated vehicles and sharing

⬆ Increase in population size or settlements   ⬇ Reduction in population size or settlements

Source: AVENUE21 after Kainrath (1997)

As already mentioned, the impact of automated vehicles on transport and urban spaces is thus by and large strongly dependent on the way that automated vehicles are used and on the associated modelling assumptions.

These mainly include:

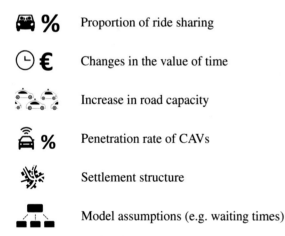

🚗%   Proportion of ride sharing

🕐€   Changes in the value of time

    Increase in road capacity

📶%   Penetration rate of CAVs

    Settlement structure

    Model assumptions (e.g. waiting times)

Moreover, the majority of modelling studies that currently exist analyse the comprehensive deployment of highly developed automated vehicles in the distant future. Possible effects in the near future, such as the potential use of automated vehicles only under certain conditions (ODD, Chap. 4.1), e.g. in specific spatial contexts (Beiker 2018: 125; Shladover 2018: 8), have rarely been the subject of scientific research to date. However, they would be of far greater importance for urban and transport planning due to the conceivably quite near-term need for action. It should be added that spatial or general effects of new transport technologies, such as those of automated vehicles, become apparent much sooner than their deployment as a result of decisions, e.g. against constructing new tram lines in Milton Keynes, UK, or in Nashville, USA (Smith 2015, The Economist 2018) being taken with automated vehicles in mind but ahead of their actual deployment.

4.3.2   CONSEQUENCES FOR URBAN BUDGETS

Alongside the analysis conducted in the project of automated vehicles' impact on transport demand and settlement structure or land use, the fiscal impacts of automated vehicles, i.e. their effect on public budgets, were also investigated using the example of Austria and specifically Vienna (Mitteregger et al. 2019).

Here it was assumed that as a result of new phenomena in the mobility system – especially the connection and automation but also electrification[1] of vehicles (in private transport) – the primary effects like possible changes in vehicle ownership, sharing, need for parking spaces, traffic efficiency and need for infrastructure would be fundamental, and would ultimately have a financial impact on public budgets (secondary effects). To analyse the fiscal impacts of connected and automated transport, the primary effects were reviewed on the basis of studies cited in the literature; the resulting secondary effects were then deduced for Austria and finally the significance of the affected revenue and expenditure categories in the budgets of the Austrian provinces and local authorities were presented with a particular focus on Vienna. The budgetary effects of a simultaneous change in PT were not taken into consideration.

OVERVIEW OF POSSIBLE FISCAL IMPACTS[2]

Figure 4.3.4 provides an overview of the possible secondary, fiscal effects due to vehicles being connected, automated and electrified, which arise as a result of the primary effects described above. In the area of infrastructure, the public sector may face considerable expenditure for the construction or adaptation of transport infrastructure because of automation and for the installation of charging infrastructure due to electrification. The creation of new or optimization of existing data infrastructure, i.e. digital transport infrastructure as a result of connection (and automation), would mean expenses for the public sector. The possible reduction in the number

Figure 4.3.4: **Phenomena in individual transport and primary and secondary (fiscal) effects through connection, automation and electrification**

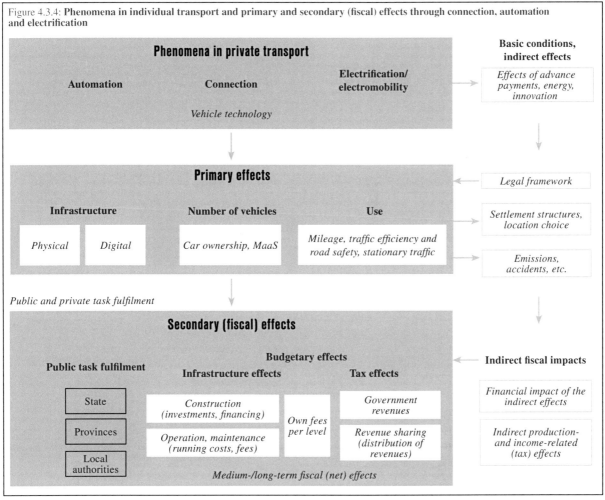

Source: AVENUE21

of vehicles due to automation might lead to reduced revenues from standard fuel consumption tax, engine-related insurance tax and vehicle tax. Furthermore, the reduced need for parking spaces associated with the smaller quantity of vehicles could also result in reduced income from management of parking spaces (parking meter charges or parking fees). In addition, the lower fuel consumption associated with increased traffic efficiency (through connection and automation) could possibly imply a decrease in fuel duty revenue, with this possibly being counteracted by the vehicles being used more (Barnes/Turkel 2017: 21). At all events, however, the electrification of vehicles alone, i.e. electrically powered vehicles, would lead to reduced income from fuel duty, as is already the case in Norway, for example (POLIS 2018: 7). Ultimately, the explicit adherence to traffic laws due to vehicles' automation, which is also relevant in terms of achieving a more efficient flow of traffic, would lead to lower income from traffic fines, such as for speeding or parking violations (Leimenstoll 2017).

In general, it is therefore clear that connected, automated and electrically powered vehicles will mainly affect transport-related revenues (i.e. public sector income) and expenditure in the areas of roadworks, road traffic and telecommunications services.

SIGNIFICANCE OF THE AFFECTED REVENUE AND EXPENDITURE CATEGORIES IN THE BUDGETS OF AUSTRIAN PROVINCES AND LOCAL AUTHORITIES

Admittedly, it was not possible to judge the exact magnitude of the secondary, fiscal effects described above – in part for want of a more detailed quantitative definition of the primary effects. However, on the basis of the qualitative description of these effects, it was possible to consider the current significance of the affected revenue and expenditure categories for the Austrian provinces and local authorities, with a special focus on Vienna, in order to identify the possible scope of these fiscal impacts. To this end, the secondary, fiscal effects depicted in italics in Figure 4.3.5 were supported with data (where possible).[3]

1   REVENUES

In Austria, the aforementioned transport-related sources of income (federal government revenues), such as fuel duty, are generally levied by the state. The provinces and local authorities then receive a share of those revenues, with the amount being defined by the distribution rule as

Figure 4.3.5: **Overview of the primary effects of connection, automation and electrification as well as the resulting possible fiscal impacts**

PHENOMENON	PRIMARY EFFECT	SECONDARY, FISCAL EFFECT	
		Income	Expenditure
Automation			*Transport infrastructure*
Connection	**Infrastructure**		*Data infrastructure (digital transport infrastructure)*
Electrification		*Fuel duty*	*Charging infrastructure, electrical grid*
Automation	**Number of vehicles** (reduction in privately owned cars)	*Standard fuel consumption tax*	
		Engine-related insurance tax	
		Vehicle tax	
Automation Connection and automation	**Use** Need for parking spaces Traffic efficiency and road safety	*Managing parking spaces (parking meter charges/fees)* *Parking/traffic fines* *Fuel duty³*	*Parking infrastructure*

Source: AVENUE21

outlined in the Fiscal Equalization Act (*Finanzausgleich-gesetz*; BMF 2018, Bröthaler et al. 2017).

Looking at the federal government revenues (Table 4.3.1), it becomes clear that transport-related revenues like fuel duty, standard fuel consumption tax, engine-related insurance tax, vehicle tax and insurance tax change as a result of shifting to connected, automated and/or electrically powered vehicles. Considering this effect is crucial bearing in mind that these revenues amounted to some €8.5 billion in 2016 or 10.7% of the total federal government revenues that year, for example. At 5.6% and 3.0% respectively, fuel duty and engine-related insurance tax constitute the largest proportion of these revenues. Over the course of time, the proportion of transport-related income in the federal government revenues is relatively stable.

Turning to the income of the provinces and local authorities after applying the distribution formula for the federal government revenues (Fig. 4.3.2), it becomes apparent

that the shares of transport-related revenues (although not earmarked) in 2017 amounted to 4.0% and 3.7% of total revenues for the provinces (excluding Vienna) and for the local authorities (excluding Vienna) respectively. For Vienna, the share of transport-related revenues totals €631 million or 4.3%. Revenues from parking meters and parking fines, which would also be subject to change as a result of connected, automated and electrically powered vehicles, constituted €115 million and €82 million respectively in 2017; taken together, this amounts to some 1.4% of Vienna's total revenue.

2 EXPENDITURE

Looking at the expenditure of the provinces (excluding Vienna) and the local authorities (excluding Vienna) in Table 4.3.2, the expenditure – which will possibly be different as a result of the advent of connected, automated and electrically powered vehicles – for roadworks and road traffic in 2017 amounted to 3.7% (provinces without

Table 4.3.1: **Amount of federal government revenues in 2007 and 2017 in millions of euros and in per cent**

FEDERAL GOVERNMENT REVENUES	2007: € MILLION	2017: € MILLION	% P.A	2017: %
Fuel duty	3,689	4,436	1.9	5.6
Standard fuel consumption tax	456	469	0.3	0.6
Engine-related insurance tax	1,410	2,389	5.4	3
Vehicle tax	1,115	38	-10.4	0
Insurance tax	993	1,128	1.3	1.4
Total transport-related revenues	6,663	8,461	2.4	10.7
Income tax	30,516	39,269	2.6	49.5
VAT	19,212	25,519	2.9	32.2
Other federal government revenues	3,870	6,015	4.5	7.6
Total revenues	*60,261*	*79,264*	*2.8*	*100*

Source: AVENUE21 after public accounts, Statistics Austria (2019a)

Table 4.3.2: **Income of the provinces and local authorities from taxes and other sources of revenue in 2017 in millions of euros and proportion of total revenue in per cent**

REVENUE 2017	IN MILLIONS OF EUROS			IN PER CENT		
	Provinces without Vienna	Local authorities without Vienna	Vienna	Provinces without Vienna	Local authorities without Vienna	Vienna
Total transport-related revenue	**1,369**	**741**	**631**	**4.0**	**3.7**	**4.3**
Profits tax	6,344	3,432	2,922	18.4	17.1	19.9
VAT	4,232	2,308	1,713	12.3	11.5	11.7
Other government revenues	1,298	1,057	793	3.8	5.3	5.4
Income tax share of federal govt. revenues	**13,244**	**7,537**	**6,059**	**38.4**	**37.7**	**41.2**
Parking meter charges incl. fees		70	115		0.3	0.8
Other local revenues	679	3,448	1,294	2.0	17.2	8.8
Total revenue	*13,923*	*11,055*	*7,468*	*40.4*	*55.2*	*50.8*
Parking/traffic fines (earmarked)	**50**	**70**	**82**	**0.1**	**0.4**	**0.6**
Other steady revenue	15,789	5,362	4,261	45.8	26.8	29.0
Revenues from assets	4,705	3,526	2,882	13.7	17.6	19.6
Total revenue	*34,466*	*20,013*	*14,693*	*100.0*	*100.0*	*100.0*

Source: AVENUE21 after public accounts, Statistics Austria (2019a)

Vienna) and 7.8% (local authorities without Vienna) of total expenditure. For Vienna, the sum is €261 million or a share of 1.8% of total expenditure. Over the past ten years, it is apparent that the share of expenditure for roads in Vienna has remained at a similar level, whereas it has slightly declined in the provinces (excluding Vienna) and in the local authorities (excluding Vienna).

Consequently, the revenue and expenditure categories affected by the introduction of connected, automated and electrically powered vehicles constitute a not insignificant proportion of total revenues and total expenditure in Austrian provinces and local authorities, especially in Vienna.

---

1 The electrification of vehicles (changing to electric propulsion systems) was included, although this is not necessarily linked to automation. The simultaneity of the two phenomena is, however, frequently emphasized in the literature (e.g. Bormann et al. 2018).

2 This section summarizes the findings of "Shared, Automated, Electric: The Fiscal Effects of the Holy Trinity" (Mitteregger et al. 2019). Expanding the analysis of CAT's possible effects to include fiscal impacts on local authorities would not have been pos-

sible without the co-authorship of Johann Bröthaler (Department of Public Finance and Infrastructure Policy at the TU Wien).

3 Here, it should be borne in mind that identifying the source of revenue is possible to a relatively precise extent, but that the revenues are not earmarked at this time. In terms of expenditure, in contrast, identifying funds is in part only possible to a rough extent (especially with regard to digitalization and electricity supply/charging infrastructure) due to budgetary categorization, the different ways in which expenditure is booked and institutional conditions (areas of responsibility). Furthermore, the empirical account is limited to the budgets of the provinces and local authorities as regional authorities (without outsourced or extrabudgetary units).

Table 4.3.3: **Expenditure of provinces and local authorities on roads and public transport in 2017 in millions of euros**

EXPENDITURE 2017	IN MILLIONS OF EUROS			IN PER CENT		
	Provinces without Vienna	Local authorities without Vienna	Vienna	Provinces without Vienna	Local authorities without Vienna	Vienna
Roads	1,283	1,783	261	3.7	7.8	1.8
Public transport	580	161	774	1.7	0.7	5.3
Other expenses	32,639	21,007	13,658	94.6	91.5	93.0
Total expenditure	*34,502*	*22,952*	*14,693*	*100.0*	*100.0*	*100.0*

Source: AVENUE21 after Statistics Austria (2019a)

Figure 4.3.6: **Impacts of automated vehicles in SAE Level 5**

TRAVEL TIME

Reduced perception of time

+
- higher productivity (carrying out different activities)
- higher driving comfort (optimized longitudinal and lateral acceleration, avoidance of driving-related stress)

−
- passengers in CAVs more likely to suffer from travel or motion sickness
- unfamiliar or lacking experience in dealing with CAVs
- shorter distances between vehicles

Reduced travel time

+
- loss of car parking traffic
- elimination of parking processes and journeys to the car park/destination

Reliable travel time

+
- almost constant speeds
- reliable and predictable routes

SAFETY

Increased traffic safety

+
- reduced driver-related causes of accidents

−
- unclear during the transition period and in mixed traffic

Accidents

-70% to -95%

Low cyber security

−
- errors in the software
- cyber attacks
- surveillance/problems with data protection

Source: BMVIT (2016b)

COSTS

Reduced operating costs

+
- more fuel-efficient and lighter vehicles
- fuel-saving vehicle styles
- lower costs for vehicle insurance
- personnel cost savings in PT
- reduced wear costs (e.g. for tyres) in PT

Higher purchase costs

+
- expensively built sensors and software components

−
- elimination of driver-related equipment inside the vehicle (steering wheel, brake and accelerator pedals)

Costs per passenger kilometre (incl. amortization)

-16% to +4%

Sinking parking costs

+
- parking in cheaper areas or areas without parking charges

Source: Bösch et al. (2017)

ROAD CAPACITY

Increased road capacity

+
- harmonization of driving modes: coordinated lane change, braking and acceleration processes
- fewer accidents
- narrower vehicles
- shorter distances between vehicles

Road capacity

+50% to +414%

Source: Van den Berg/Verhoef (2016), Bösch (2016)

NEW USER GROUPS

Tapping into new user groups

+
- lower demands on drivers/passengers (no driving licence necessary)

People with limited mobility in Austria

10% to 38%

Source: Sammer et al. (2013)

ENERGY AND THE ENVIRONMENT

Energy efficiency (fuel savings)	
-11% to -47%	

Emissions	
CO	-32% to -61%
CO_2	-61% to +105%
NO_x	-2% to -18%

There are occasional studies concerning automated vehicles' impact on energy and the environment. However, even here the effects vary widely. Further research is required, especially in terms of the effects on traffic and mobility.

Source: Fagnant et al. (2015), Milakis et al. (2017), Wadud et al. (2016)

ECONOMY AND PUBLIC FINANCES

Profit per vehicle	€2,960 to €3,900
Value added	+30%

Loss of taxes/revenue	Current significance in Austria
Standardized consumption tax	€2,960 to €3,900
Engine-related insurance tax	+30%
Parking meter charges	€2,960 to €3,900
Parking/traffic fines	+30%

There have been barely any studies in this area. At most, studies demonstrate the current significance of the possible loss of income sources for the public sector; just how great this loss will be is, however, largely uncertain and dependent on the impacts described above.

Source: Clements/Kockelman (2017), Fagnant/Kockelman (2015), Mitteregger et al. (2019)

4.4

AUTOMATED DRIVABILITY: A NUANCED PICTURE OF THE SPATIAL DEPLOYMENT OF CONNECTED AND AUTOMATED VEHICLES

Insight into the dissertation of Aggelos Soteropoulos

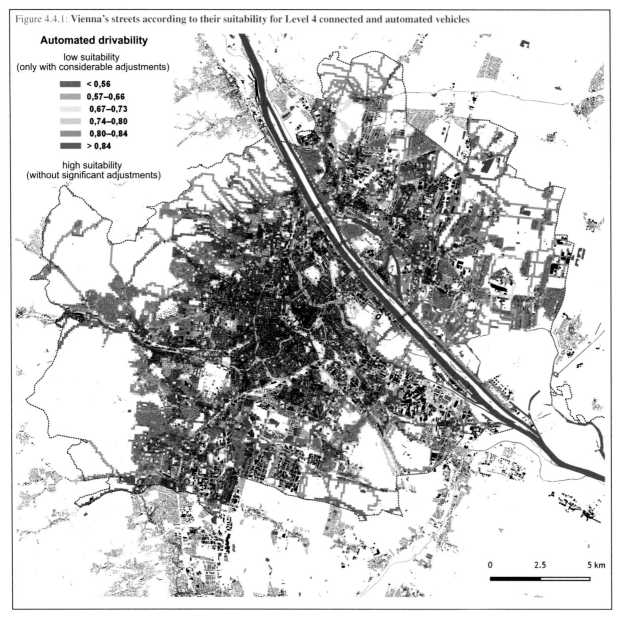

Figure 4.4.1: **Vienna's streets according to their suitability for Level 4 connected and automated vehicles**

Automated drivability

low suitability
(only with considerable adjustments)

- < 0,56
- 0,57–0,66
- 0,67–0,73
- 0,74–0,80
- 0,80–0,84
- > 0,84

high suitability
(without significant adjustments)

0   2.5   5 km

Source: AVENUE21

As explained in Chapter 4.3, simulations and forecasts on the use of CAT only take into account the qualitative differences between streetscapes to a limited extent. Thus, it is assumed that e.g. motorways or special areas constitute "easier" automation tasks and consequently automated driving systems (SAE International 2018 does not yet define any connection here) will be used there sooner (see also Chap. 4.1). However, beyond this broad comparison, a nuanced view that includes the characteristics of the streetscapes and environmental conditions in these two categories is lacking. The term operational design domain (ODD) was introduced

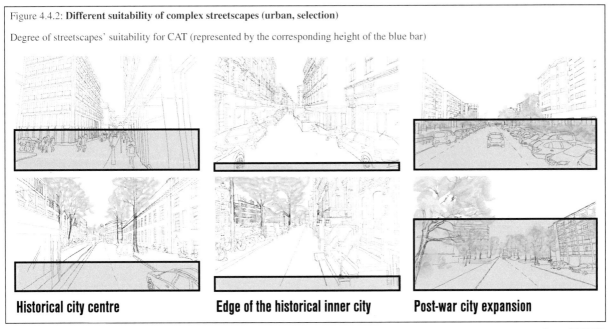

Figure 4.4.2: **Different suitability of complex streetscapes (urban, selection)**

Degree of streetscapes' suitability for CAT (represented by the corresponding height of the blue bar)

Historical city centre   **Edge of the historical inner city**   **Post-war city expansion**

Source: AVENUE21

to the famous SAE levels with their revision in 2016 to convey the range of complexity in environments for vehicle automation (SAE International 2018: 14). For example, an automated driving system can only function at low speeds, in good weather and during the day (Fraade-Blanar et al. 2018: 13). These limitations would correspond to the ODD for which the system was designed.

In the AVENUE21 research project, a first step towards developing an index of automated drivability was taken in order to take into consideration in a more nuanced way the demands on automated driving systems that depend directly on the complexity of the ODD (see Fig. 4.4.3). The starting point for this index was the fact that certain road contexts increase the demands on automated driving systems (Metz 2018: 3). This results in particular from the way they function: automated driving systems have to perceive their surroundings using various sensors, map it using the sensor data (perception and cognition), make the corresponding driving decisions (planning and control) and communicate them to the passengers and other road users (human-machine interaction; Ritz 2018: 41). Different deployment environments and associated conditions can complicate these processes, for example when a large number of different road users have to be recognized and their future movements predicted or anticipated (Shladover 2018: 31).

EUROPEAN CITIES: VARIETY OF STREETSCAPES

Especially in European cities, there is a wide variety of streetscapes, which differ greatly in terms of their structural and infrastructural appearance, the surrounding architecture and the range of road users and hence make very different demands on automated driving systems (see Fig. 4.4.2). Streetscapes also have different functions (connection or access function) and congestion levels or traffic density (FGSV 2006: 8; Marshall 2005: 50). As points of connection between roads, junctions are also diverse (right of way for traffic coming from the right, priority traffic signs, light signals, roundabouts, etc.) and depend on the characteristics of the connecting roads (FGSV 2006: 54).

Figure 4.4.3: **Basic relationship between the automated driving system and the ODD**

On streets with lower complexity, automated driving systems can predict the processes they need to carry out more precisely and more easily (left). In contrast, streets with high complexity complicate the precision of this prediction (right).

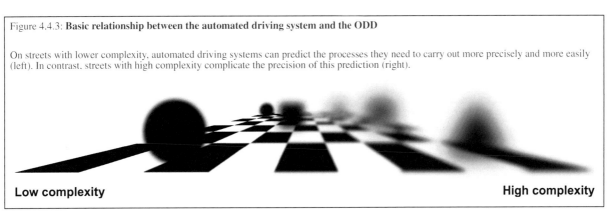

Low complexity   **High complexity**

Source: AVENUE21

FACTORS IN STREETSCAPES' SUITABILITY FOR THE USE OF CONNECTED AND AUTOMATED VEHICLES: AUTOMATED DRIVABILITY

Proceeding from these points, factors were developed that identify the suitability of streets or roads for the use of CAVs – i.e. their automated drivability – and hence also define the spatially nuanced use of CAT. These factors comprise (see Fig. 4.4.4):

1 the number of objects in the streetscape,

2 the variety of objects in the streetscape,

3 the stability of the ODD,

4 the permitted speed range and

5 the condition of the infrastructure.

ASSESSMENT OF AUTOMATED DRIVABILITY: VIENNA AS A CASE STUDY

Using the city of Vienna as a case study, example criteria were derived and combined in light of the developed factors and dependent on the availability of data; these criteria map as well as possible the developed factors and hence streetscapes' suitability for the functional deployment of automated driving systems (see Fig. 4.4.5). The criteria used are an attempt to convey the heterogeneity of various streetscapes in their location in the city. Figure 4.4.1 shows the assessment of automated drivability using the example of the city of Vienna. It makes clear that cities are by no means homogeneous spaces when it comes to the possibilities of deploying CAVs. Rather, the derived criteria reveal that a spatially selective deployment of CAVs is probable due to the differing suitability of streetscapes in cities and that the advantages and disadvantages of the deployment of CAVs in the long transition period will be distributed unequally.

One consequence of this heterogeneity is a shift in locations' popularity due to the accessibility with CAVs of passenger and freight transport. It is noticeable that more car-friendly peripheral areas of the city (which are shown in green in Fig. 4.4.1) tend to have an advantage over older, largely more complex neighbourhoods in the city centre. While the former could be accessed by CAVs without (considerable) adjustments and presumably relatively soon, the deployment of CAVs in the latter would only be conceivable if considerable adjustments were made to the streetscape (built infrastructure), speeds were generally reduced and the area were upgraded with digital infrastructure (shown in red) – yet for such changes there are no standards whatsoever at this time. Furthermore, it is clear that European

Figure 4.4.4: **Short description of the developed factors**

FACTOR	SHORT DESCRIPTION
Number of objects in the streetscape	<ul><li>Existing road users and other movable objects (e.g. animals) and static objects (e.g. road signs, traffic lights, road markings)</li><li>Frequent reason for problems and incidents during tests (Favarò et al. 2018: 142)</li><li>Objects in the streetscape like parking cars, street lights, road signs or advertising as barriers to perception and cognition</li></ul>
Variety of objects in the streetscape	<ul><li>Heterogeneity of road users (e.g. pedestrians, cyclists) and other objects (e.g. traffic lights at railway crossings)</li></ul>
Stability of the ODD	<ul><li>Constancy of the environmental conditions</li><li>Complicated (limited) conditions for the cameras' and sensors' functionality due to rapidly changing static obstacles (e.g. frequent changes in road signs, construction work) and rapidly changing weather conditions or vegetation</li></ul>
Permitted speed range	<ul><li>Higher speeds require the automated driving system to have a shorter reaction time (e.g. time for sensors to perceive the surroundings and for the software to process the data; Campbell et al. 2010: 4664)</li></ul>
Condition of the infrastructure	<ul><li>Existence of infrastructure (like road markings) as well as their condition or quality</li><li>Perception and cognition of road markings in bad condition, lacking delineator posts or potholes and different colour values, for example as a result of patch-ups or comparisons between asphalt and concrete (state of the lane) as a problem for sensors and software (Fellendorf 2018: 5; Alkim 2018: 22)</li></ul>

Source: AVENUE21

Figure 4.4.5: **Overview of the derived criteria for mapping automated drivability using the example of Vienna (excerpt)**

NUMBER OF OBJECTS IN THE STREETSCAPE

 Existence of a pedestrian zone or shared space

 Overwhelmingly residential area type

 Average lane width

NUMBER AND VARIETY OF OBJECTS IN THE STREETSCAPE

 Number of PT stops

 Number of traffic lights

VARIETY OF OBJECTS IN THE STREETSCAPE

 Existence of cycling infrastructure in the lane

PERMITTED SPEED RANGE

 Reduced speed in the section of the road

STABILITY OF THE OPERATIONAL DESIGN DOMAIN

 Existence of vegetation (trees, bushes …)

CONDITION OF THE INFRASTRUCTURE

 Difference in road surfaces

Source: AVENUE21

cities, which frequently have historical, pedestrian-oriented centres with narrow streets, cannot accommodate motorized and connected transport. Transportation network companies like Uber or Lyft are now reacting to this by moving their services to two-wheeled vehicles in such areas.

This assessment for Vienna has provided an initial overview of the suitability of road environments for the deployment of CAVs. With more data, the analysis can be differentiated. Nevertheless, in light of the developed index, the mapping of Vienna's automated drivability (Fig. 4.4.1) offers an insight into the areas that can be accessed by CAT without (considerable) adjustments and that are in no way equally distributed across the city. In addition, a social space analysis could identify which social groups live on the green or red-marked streets and what mobility styles are dominant there. This knowledge is of fundamental importance for policy and planning in order to develop nuanced measures.

4.5

TRANSITION MANAGEMENT IN PIONEERING REGIONS AROUND THE GLOBE

In the context of the effects of CAM on cities and urban regions, so-called pioneering regions will be described. This means cities or urban regions that are actively researching, planning and developing as well as testing and demonstrating CAVs and CAM and that are pushing to implement CAM in view of regional challenges.

The focus of this chapter is on the challenges that cities and urban regions hope to address by deploying CAVs, and the transition initiatives that make cities and urban regions co-producers in the development of CAVs.

SELECTION OF THE PIONEERING REGIONS

In the course of our research, some 20 urban regions were chosen, analysed and compared. After helpful discussions with experts, five of them were selected as pioneering regions: they lie in the USA (San Francisco), in the UK (the metropolitan region of London), in Sweden (Gothenburg) and in Japan (the metropolitan region of Tokyo). The fifth pioneering region chosen was the city state of Singapore (see Fig. 4.5.1). The aim of the study was to review as broad a spectrum of different reference regions as possible which reveal relevant visions of the future and development possibilities in which transition

initiatives are visible in strategies and programmes and which have already found recognition for their contributions to the global discourse around CAT.

THE FOUR ANALYTICAL CATEGORIES

To analyse these pioneering regions, the theoretical frameworks of transition theory (Rotmans et al. 2001, Kemp/Loorbach 2003, Geels 2005) and the multi-level perspective (MLP; Geels 2010) were applied. The heuristic methods of MLP view system transitions as relational processes which, through the interaction between the developments, result in three dimensions (Rotmans et al. 2001):

- The dimension "landscape" covers local or translocal aspects of a city/region, such as the effects of demographic change.

- The dimension "regime" includes a country's prevalent practices, regulations, etc.

- The dimension "niche" encompasses "protected spaces" like research laboratories, start-ups, etc., which experiment with innovations.

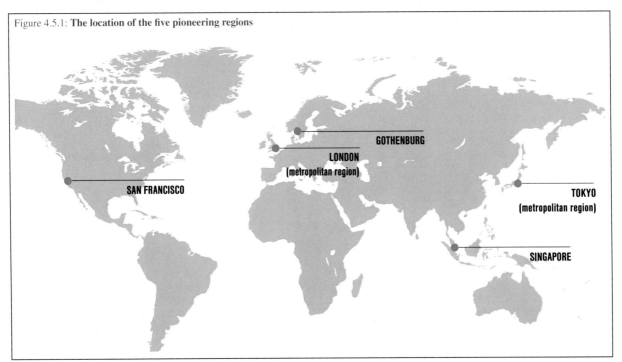

Figure 4.5.1: **The location of the five pioneering regions**

GOTHENBURG

LONDON
(metropolitan region)

SAN FRANCISCO

TOKYO
(metropolitan region)

SINGAPORE

Based on these two theories, four analytical criteria were developed for the analysis and comparative evaluation of the pioneering regions:

1 "Drivers" refer on the one hand to the pressure to act resulting from the social and ecological challenges ("landscape"), and on the other to the pressure from the "niche".

2 "Innovation networks" refer to the key actors and alliances involved, which jointly plan and initiate progress and change ("regime" and "niche").

3 "Development narratives" refer to visions and models that arise in corporative and political negotiation processes (see Chap. 4.6).

4 "Transition initiatives" are associated strategies and programmes that are intended to provide orientation for joint action (e.g. in the form of road maps).

Research questions were formulated for each of the four analytical categories. With all of these questions, the focus was on the interplay of CAM strategies and the long-term plans for urban development and transport in the respective pioneering region (see Fig. 4.5.2).

SUMMARY OF THE RESULTS: CONCEPTION OF CONNECTED AND AUTOMATED MOBILITY FROM THE PERSPECTIVE OF CITIES/URBAN REGIONS

The approaches taken by the five selected pioneering regions (cities/urban regions) are summarized below with reference to the four analytical categories: (1) drivers, (2) innovation networks, (3) development narratives and (4) transition initiatives.

■ SAN FRANCISCO (USA)

Drivers: The current urban challenges in San Francisco are an acute housing shortage, a fragmented public transport system, unequal growth patterns between the districts, a growing evening economy and limited mobility options, especially for low-income households, seniors and people with disabilities (SFMTA 2016a, 2016b: 1; SFMTA 2018). The tech giants of Silicon Valley in particular are eager to be allowed to test their technologies on the streets of San Francisco.

Innovation networks: The San Francisco Municipal Transportation Agency (SFMTA) is the main actor supporting automated transport in San Francisco. A meeting place, an agency and an idea incubator called Smart City Institute (also known as Superpublic) have been set up (ATCMTD, City of San Francisco 2016, SEE 14). With the "Community-Guided Engagement Plan", a framework has been created to bring about different kinds of cooperation and alliance between various sectors and actors (i.e. cross-sector collaboration). In addition, the city has entered into an alliance with the World Economic Forum with a special focus on CAM.

Development narratives: The first two CAM narratives mainly refer to safety ("Vision Zero SF") and improved mobility options for low-income households ("Transportation is the greatest equalizer of all"). CAM is addressed directly by the third narrative, which conceptualizes a medium- and long-term vision of the future that revolves around shared, electric, connected and automated vehicles (SECAVs).

Transition initiatives: The home of the sharing economy has set itself the task of introducing, testing and later incorporating a large quantity of shared forms of mobility in the wide range of mobility services available. In order to achieve this long-term goal, there is considerable reliance on the deployment of CAVs. The transition to SE-

Figure 4.5.2: **Categories and main research questions in the context of analysing the pioneering regions**

ANALYTICAL CATEGORIES	RESEARCH QUESTIONS
Drivers: *challenges and expectations*	• What induces cities/regions (policymakers, administration) to invest in CAM? • What drivers emerge due to the pressure to act from the "landscape" or "niche"?
Innovation networks: *actors and cooperation*	• Who are the relevant actors in a city/urban region who will advocate the introduction and distribution of CAVs? • What types of cooperation are there? What alliances have been formed? • What steps have been taken to achieve progress and change towards CAM?
Development narratives: *visions and models*	• What visions of the future have been jointly formulated? What strategic goals and models have been derived from them? • What narratives are preferred to make CAM attractive ("basket of images")?
Transition initiatives: *strategies and programmes*	• How is the possible transition to CAM envisioned? What corresponding aims and strategies are there? What investment programmes have been agreed? • How have the strategies and implementations been formulated (e.g. as road maps)?

CAVs was conceived in the form of a step-by-step plan. The most important medium-term aim is to reclaim space in the inner city (until now used for parking spaces) in order to create e.g. more housing through densification. In no other region of the world have so many real and virtual kilometres been driven in automated transport than in California (US Senate Hearing 2018).

■ LONDON AND METROPOLITAN REGION (UK)

Drivers: The current challenges facing urban development in London are the rapidly growing population in the region, increasing traffic jams on the main transport routes, rising social inequality and the acute need to compensate for rapid urban growth by means of decentralization (GLA 2015). The widely shared expectation that the development and deployment of CAVs will bring technical and economic benefits is the most important driver of investments in the research and application of CAVs in the United Kingdom.

Innovation networks: In the UK, there are numerous university institutes, influential non-university research organizations and think tanks that contribute greatly to the local and national discourse and to the promotion and development of CAVs. The British Department for Transport (DfT) and Department for Business, Energy & Industrial Strategy (BEIS) together launched the Centre for Connected and Autonomous Vehicles (CCAV) in 2017 (UK Parliament 2017). The CCAV is the single point of contact for the development of CAT in the UK.

Development narratives: The most important development narratives and strategies for the future of mobility and the visions for CAM embedded within that are concentrated on economic competitive advantages, on health and well-being, on social inclusion, on "good" growth and on the notion of a polycentric city (GLA 2017, NLA 2017, TfL 2017).

Transition initiatives: London's long-term urban development strategy is based on two pillars. There is a focus on the one hand on decentralized growth and on the other on citizens' health and well-being (TfL 2017). A detailed road map for the development of CAT in the UK was formulated after extensive public consultations (DfT 2015). In the course of this, a special process for selecting sponsorship projects to develop and test CAVs was implemented. Numerous projects have studied how citizens view and deal with CAM in everyday life.

■ GOTHENBURG (SWEDEN)

Drivers: The challenges facing Gothenburg are a growing population, rising economic and social inequality, increasing freight traffic (the largest port in Scandinavia is in Gothenburg), the effects of climate change on numerous urban districts (rising sea level) and an expanding knowledge-based economy which needs new kinds of socio-economic communication spaces. The presence of a range of companies (like the Volvo Group) has raised expectations in the city that it can gain competitive advantages by investing in the research and development of CAVs.

Innovation networks: The cooperation and collaboration platform Drive Sweden is the most important organization to be set up in Sweden to publicly support the development of CAVs (Drive Sweden 2018). With so-called "Co-Creative Labs" (SAFER 2017), the urban planners in Gothenburg are attempting to involve the population in the joint development of a citizen-driven perspective for CAM.

Development narratives: The main urban development narrative in Gothenburg is the "Compact City" (City Planning Authority 2014). Its aim is on the one hand urban densification and on the other developing and strengthening Gothenburg as a hub in the knowledge-based economy. Awareness is being raised in the population for a change in urban mobility and transport ("A completely new type of mobility").

Transition initiatives: Gothenburg is intended to become a city of short distances ("closely-connected city"; Urban Transport Committee 2014). Drive Sweden sponsors various cooperative research and development projects like the deployment of self-driving shuttle buses that are suitable for a compact city. An interactive and evolving road map for the development of CAM in Sweden on the Drive Sweden website informs interested actors about the development status and possible stages of the transition to fully automated driving. A big step for the city of Gothenburg and its automotive and communication sectors was the construction of Asta Zero, a CAV testing ground of international significance.

■ TOKYO AND METROPOLITAN REGION (JAPAN)

Drivers: The challenges in Tokyo (and Japan as a whole) are a dramatically shrinking and ageing population, an acute shortage of labour and stagnant economic growth (Funabashi 2018, IPSS 2017). Furthermore, hosting the Olympic Games that were originally planned for 2020 was and is a driver for the rapid development of CAVs.

Innovation networks: With the SIP ("Cross-Ministerial Strategic Innovation Promotion Program") – and the support of the prime minister – the Council for Science, Technology and Innovation created a unique kind of horizontal, cross-departmental organizational structure (Amano/Uchimura 2016, 2018; SIP 2017). This started a process to overcome the compartmentalization of institutional "silos". A specific area of this programme, the SIP-ADUS (Automated Driving for Universal Services), enables both the coordination of research and development (SIP-ADUS for CAM) and the orchestration of the "Triple Helix" in Japan between government actors (cross-departmental), industry and research institutes (Amano/Uchimura 2016, 2018; SIP-ADUS 2016).

Development narratives: With "Mobility bringing everyone a smile" (Kuzumaki 2017), the city government wants above all to include seniors. In contrast, the two narratives "World's safest and smoothest road traffic system" (Prime Minister's Cabinet 2017) and "World's most advanced IT nation" are aimed at businesses. The road map "Society 5.0" refers to the current public discussion in Japan about the future of society in the age of digitalization, robots and artificial intelligence (FIRST 2018). In Japan, the safety aspect concerns the constantly growing threat of major earthquakes; for this reason, special safety systems have already been installed in public transport.

Transition initiatives: Tokyo's development up to 2020 was shaped by several action plans. The main idea was to equip the megalopolis of Tokyo with numerous compact city hubs (MLIT 2013, Tokyo Metropolitan Government 2016). The two most important social issues for which CAM is expected to provide solutions are firstly the ageing population and secondly the shrinking workforce. Most CAM projects currently aim to connect Tokyo with its suburbs and surrounding rural area (Prime Minister's Cabinet 2017). Under the coordinating aegis of the cross-departmental SIP-ADUS, Tokyo has produced the most detailed CAM road map of all the pioneering regions discussed here. It contains all the aims, strategies and concepts for the deployment of connected and automated vehicles, including a timetable for their implementation and roll-out in Japan by the year 2030. A unique feature of the strategy in Japan is the successive creation of a dynamic, nationwide, digital map of the streetscapes (Koyama 2015).

The ageing population and the accompanying dramatic reduction in the rural population are among the dominant issues in Japan. In the long-term vision "The Grand Design 2050" (MLIT 2013), a densification of smaller cities and towns in suburban and rural Japan is proposed (Funabashi 2018). Most CAM projects currently focus on these matters.

■ **SINGAPORE**

Drivers: Urban challenges in Singapore are the acute shortage of land in combination with the growing population and their increasing demands for better mobility options, the construction and expansion of urban areas ("New Towns") and a lack of skilled workers (LTA 2013, 2018; Loo 2017; Human Resources 2018). Other drivers are improving global competitiveness and maintaining location quality.

Innovation networks: The cross-ministerial Committee on Autonomous Road Transport for Singapore (CARTS) was created as part of the city administration of Singapore to focus exclusively on testing and implementing CAVs and CAM in Singapore (Huiling/ Goh 2017, MOT 2017). The committee is accompanied by three working groups (cross-sector and transdisciplinary), which are responsible for drawing up medium- and long-term visions for CAM in Singapore and checking the parameters for possible applications (costs, regulatory framework, etc.).

Development narratives: The three most important development narratives are Singapore's all-encompassing digitalization strategy "Smart Nation", the vision of a city without cars "Car-lite City" (LTA 2018) and the model of a "Livable City" (CLC 2014, Smart Nation Singapore 2018). The innovation discourse in Singapore is shaped by the possible compatibility of new technologies with an identity-forming and "emotional" urbanism ("Can you love a smart city?"). In order to limit the number of authorized cars, since February 2018 new cars are only permitted in Singapore if they replace an existing car.

Transition initiatives: Compared with all the other pioneering regions, in Singapore CAVs enjoy the highest level of integration in urban development. The centralized government of the city state of Singapore has pursued the ideal of integrated planning for 50 years by means of various instruments of interinstitutional collaboration (URA 2016). On the basis of intensive cross-sector cooperation (including cooperation with investors), locations and use cases for CAVs can be planned and implemented in an appropriate and targeted manner. Two new high-density settlement areas ("New Towns"), which should be completed by 2024, are already designated for diverse location-specific applications of CAVs for both peak and off-peak times and for both daytime and night-time operation (see Fig. 4.5.3). In addition, intensive work is underway in Singapore on education formats and programmes to qualify and train workers in the field of CAM.

CONCLUSION: "AUTOMATED AND CONNECTED MOBILITY FOLLOWS (INNOVATION) NARRATION"

The results of the study suggest that despite the increase in safety, traffic efficiency, economic advantages or savings on parking spaces generally being the most frequently cited arguments for the deployment of CAVs in the city, the local designs and CAM strategies can in fact be surprisingly diverse. This diversity even reveals counternarratives to the current economic mainstream. The case studies show that technology is not deterministic for CAVs and CAM and that such strategies are the result of complex negotiation processes. It is also

Figure 4.5.3: **Connected and automated vehicles in daytime and night-time operation, concepts for the "New Towns" in Singapore**

important to recognize that all cities are preparing for a system-wide transition period of several decades. However, most of them have identified specific urban spaces where they can implement CAVs of lower automation levels much earlier.

The four analytical categories (1) drivers, (2) innovation networks, (3) development narratives and (4) transition initiatives show in all five pioneering regions that the intensity and nature of the local innovation discourse is decisive for the complex processes involved in the transition to CAM-based solutions. It should also be noted that the introduction and distribution of CAVs is only a part of the digital transformation of all aspects of life ("fourth industrial revolution"). Considering the wide-ranging consequences of the digital transformation, the crucial issue for the deployment of CAVs is therefore not whether the city is ready for CAVs or whether CAVs are ready to be deployed in the city, but rather how the actors in the city lead a discussion – themselves and in cooperation with others – about sustainable urban development in the digital age (co-production in networks). The real-world implementation of CAM then follows the narrative that was established in the discourse.

4.6

NEGOTIATING A DOMINANT NARRATIVE ABOUT CONNECTED AND AUTOMATED MOBILITY IN EUROPE

Insight into the dissertation of Andrea Stickler

The role that automation will play in the transport system is not only defined by technological possibilities but is largely dependent on which discourses and narratives[1] about CAM become established in the public mindset. These visions of the future and hence the acceptance of CAVs are greatly influenced by the messages conveyed by various branches of the economy, the media, research, film or art and indeed politics (Diehl/Diehl 2018, Manderscheid 2018, Berscheid 2014). However, the political discourse plays a unique role because it has taken an extremely controversial turn in recent years, is leading or has led to certain political actions and hence has a determining influence on the conditions of the technology's deployment (through circulating knowledge, test beds and examples of best practice, support schemes, regulations, etc.). For this reason, the question arises in what direction politics (actors, institutions, processes and issues) will move in the course of CAVs' introduction and what policy and planning framework will be defined that will ultimately affect future steering options.[2]

Although the hopes for technological innovations in transportation have been evident in the political discourse for many years (e.g. European Union 2011), the expectations of the positive effects of CAVs are relatively new. The transport ministries of the G7 states first published a declaration on the strategic promotion of CAM in 2015. At EU level, a consensus between the 28 transport ministries on the joint promotion of CAVs' introduction was formulated with the Declaration of Amsterdam in April 2016 and called on the European Commission to develop a European CAM strategy (European Commission 2016). In the context of this confidence around the implementation of connected and automated mobility and around the anticipated positive effects, certain narratives have developed about how CAVs should be deployed in Europe and the effect they will have.

In the area of transport policy, the political and discursive arena is currently characterized by competing ideas, interpretations of problems, and interests, as well as by the various stakeholders' different communication and power resources. What is considered verified knowledge about CAM is closely connected to questions of power, because co-determining the ability and assertiveness of a dominant narrative at political level differs enormously between the various stakeholders. Following the multiple-stream approach (Kingdon 1986), not only the power structure but also the perceived pressure to act and existing transport policies, solutions and infrastructures are relevant for setting political narratives and agendas in a specific time frame (see Fig. 4.6.1).

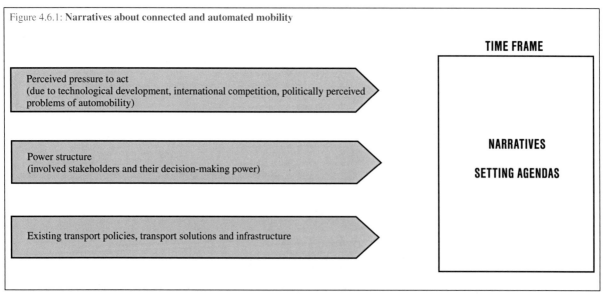

Figure 4.6.1: **Narratives about connected and automated mobility**

TIME FRAME

Perceived pressure to act
(due to technological development, international competition, politically perceived problems of automobility)

Power structure
(involved stakeholders and their decision-making power)

Existing transport policies, transport solutions and infrastructure

NARRATIVES

SETTING AGENDAS

Source: AVENUE21 after Kingdon (1986)

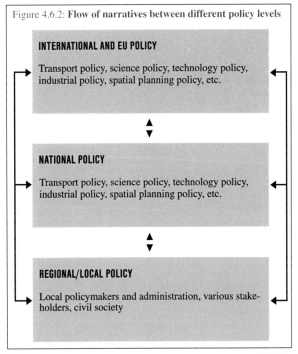

Figure 4.6.2: **Flow of narratives between different policy levels**

INTERNATIONAL AND EU POLICY

Transport policy, science policy, technology policy, industrial policy, spatial planning policy, etc.

NATIONAL POLICY

Transport policy, science policy, technology policy, industrial policy, spatial planning policy, etc.

REGIONAL/LOCAL POLICY

Local policymakers and administration, various stakeholders, civil society

Source: AVENUE21

In the area of transport policy, the non-existent sustainability of current car traffic is largely undisputed, which is why the discourse on CAM within the EU increasingly revolves around environmental policy argumentation patterns. An ecological modernization (Bemmann et al. 2014) of cars should help to tap into the positive synergetic effects of vehicle automation, electrification and connectivity while simultaneously strengthening the competitiveness of the economy (and automotive industry) as well as improving European integration. As such, the narrative of a transformation in automobility with CAVs by means of an infrastructure prioritization of important transport routes within the EU (here in particular the C-ITS corridor from Rotterdam to Frankfurt and Vienna) is one that overwhelmingly has faith in technology and is growth oriented.

The EU's narrative on CAM, which is conveyed via policy programmes, political panels, events and the media, is becoming the new way to think and act in various areas and institutions (which is why the flow of narratives between different policy levels and policy fields should be considered, see Fig. 4.6.2).

In political debates on CAM, undeniably relevant social challenges like climate protection, safety, the space consumed by cars and opportunities for people with restricted mobility are addressed and raised in public. However, design work is applied to these roblems in this context, for which specific technological solutions are then sought and developed in a cooperation between politics, various industries and research institutes. Hopes for the future and breaks with the past form a significant basis for the negotiation of a dominant narrative about CAM in this context. Predominantly with regard to the significance of automobility, the political discourse today is still sharply divided; even historically, there have been alternating phases of advocacy and criticism of automobility (Paterson 2007).

Narratives at EU level contrast with other (inter)national notions of CAM, exert pressure on various political levels and are partly adopted and "translated" by national or local institutions (Clarke et al. 2015). This translation allows an entirely new negotiation of the issue, with new tensions, risks and opportunities. Certain aspects might be emphasized or ignored, and unexpected effects may emerge as a result of path dependencies, lock-in effects, challenges and tensions (see Fig. 4.6.3). For this reason, it is necessary to include specific spatial circumstances in considerations in order to understand the relationships between the political narratives and certain policy and planning actions (Jäger 2015: 112–113). Only by doing so does it become clear that although political discourses and narratives will have an impact on the

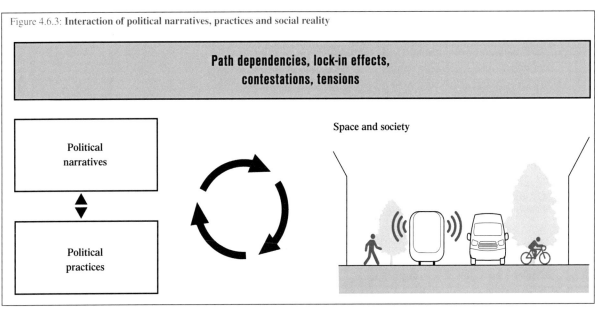

Figure 4.6.3: **Interaction of political narratives, practices and social reality**

Path dependencies, lock-in effects, contestations, tensions

Political narratives

Political practices

Space and society

Source: AVENUE21

new transportation revolution, they will not dictate the revolution because the transformation of automobility is embedded in a complex social and politico-economic nexus that has evolved over time (Urry 2004, Paterson 2007). Nevertheless, political discourses can contribute greatly to the new transportation revolution as they are linked with certain discourse practices that can indeed have an influence on structure and space.

The technical and growth-oriented notion of CAM within the EU is already being called into question. Particularly at local/municipal level, the negative effects of automobility and the limits to its growth are tangible, which is why further traffic growth with CAVs is the subject of very contentious debate. Doubts can also be raised about a growth-oriented transformation narrative from an empirical and theoretical perspective: statistics point to a considerable increase in road traffic (Umweltbundesamt 2018, Statistics Austria 2018), which will soon reach its (ecological) limits. Social processes like individualization, flexibilization and social acceleration (Rosa 2013, Honneth 2016) as well as the dynamics of the global economic system (Boltanski/Chiapello 2001, Schwedes 2017) tend to thwart rather than encourage environmentally friendly mobility at both a cultural and structural level. In light of this trend towards increasing traffic and a growth-oriented narrative around CAT, especially in cities many adverse effects of automobility (use of space, negative impact on pedestrian and bicycle traffic, sound pollution, etc.) could be exacerbated or maintained.

Despite this, in some discourses the potential of a radical transformation with CAVs is also recognized, because progress in the connection and automation of vehicle technology may change not only the car as a piece of technical equipment, but also the car's significance as a private commodity and ultimately how and why individuals are mobile (Kellerman 2018, Canzler/Knie 2016). However, these culture- and growth-critical discourses are often neglected in political debates at European level. To increase the politicization of CAM, more debate would be necessary between predominant and alternative narratives. Yet this would require the political level to think beyond an ecological modernization of the car, to discuss opposing visions of the future with CAVs and to hold more talks in the sociopolitical arena about the various possibilities for implementing CAVs.

In summary, it can be stated that especially due to the competing interests and interpretations of problems in modern-day transport policy, current political narratives around CAM should be critically challenged, because:

■ In the EU, a certain narrative is evolving between actors' conflicting priorities regarding how CAVs should best be deployed. This is leading to altered legal frameworks and funding

strategies as political practices that will have a considerable impact on the future deployment of CAVs.

■ The EU's narratives and political practices acquire a different meaning at local level, where they are being called into question and renegotiated. This negotiation process should be strengthened intentionally and opened up to various interest groups.

■ Research can help expose the prevailing power relations that shape the discourse around CAM and serve as "early warning systems" for potentially unexpected consequences.

—

1   The term "narrative" is interpreted here as the communicated and staged narrative or myth around CAVs and CAM. These narratives refer to control systems that give rise to certain forms of knowledge and meanings and hence structure the public and/or political discourse (Viehöver 2001: 177–178).

2   Specifically, this addresses the question of how CAM is uttered as an object of knowledge by various stakeholders. We take a discourse theory approach and refer methodically to discourse theories such as those of Keller (2004) and Jäger (2015).

4.7

PLANNING APPROACHES THAT PROACTIVELY SHAPE URBAN FUTURES WITH CONNECTED AND AUTOMATED VEHICLES

Insight into the dissertation of Emilia Bruck

Even though connected and automated mobility is increasingly being discussed in light of existing urban development aims (Heinrichs et al. 2019), suitable planning approaches to steer the urban integration of CAVs have not yet been adequately addressed (Guerra 2016, Fraedrich et al. 2018). With few exceptions (Chap. 4.5), urban planners have not had sufficient knowledge of the consequences of CAM on urban planning and settlement structure to be able to establish development priorities and manage investments appropriately (Fraedrich et al. 2018).

There are uncertainties regarding impacts on infrastructure, urban space and society because it has not yet been possible to analyse the local benefits of CAVs in terms of environmentally and city-friendly mobility. However, when one considers that the nature and extent of CAM's possible effects will in part only be identifiable after its implementation (Guerra 2016), urban planners need to reflectively engage and be actively involved in shaping that implementation in good time in order to act as a driving force and represent the city region's issues concerning technology development, science funding and policy. Assuming that CAVs will not be deployed spatially and socially from the outset, but only (co-)generated by public and political negotiation (Guthrie/Dutton 1992, see also Chap. 4.6), urban planners also bear creative responsibility, as the spatial objectives set during the formative phase will shape and possibly have a lasting impact on the further development of the technology. Now the urgent question presents itself as to at what point specific planning measures should be adopted and what shape the necessary vision and strategy-finding process should take.

Figure 4.7.1: **City dashboard**

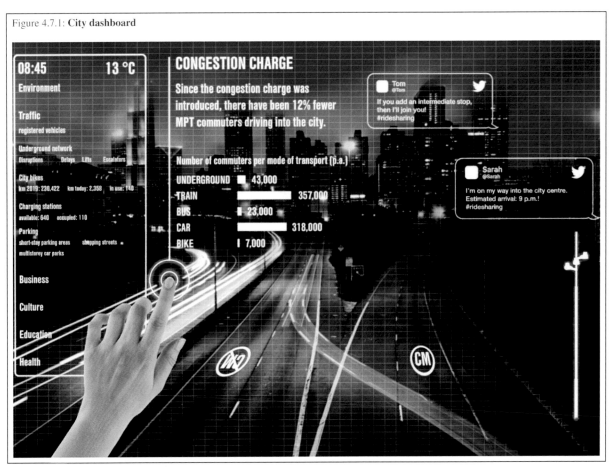

4.7.1 TECHNOLOGICAL CHANGE AS A CHALLENGE FOR URBAN PLANNING

The introduction of CAVs fits into a process of upgrading information technology in numerous European (London, Vienna, Amsterdam, Barcelona, Dublin, etc.) and international (New York, Singapore, Hong Kong, Pune, etc.) cities. Over the past two decades, strategy papers have been written on this topic with the "smart city" as their leitmotif; although they differ in their specific objectives, development processes and implementation strategies, they largely aim to adapt technological and economic trends to local development requirements (Townsend/Lorimer 2015). Given limited infrastructure capacity, fiscal austerity, the climate crisis and ongoing urbanization trends, it is hoped that smart infrastructure development will guarantee urban resilience and sustainability (White 2016). With the aspiration to monitor, manage and steer urban processes in real time (Kitchin 2015), a technology and management-oriented planning approach is taking hold which is based on the conviction that with "better" data and "better" models, uncertainties and risks will be considerably reduced if not entirely eliminated (Hillier 2016: 300). The prediction and control of events within infrastructure networks or public spaces (see Fig. 4.7.1) is at the heart of this promise (Picon 2015). Hence the idealized smart city developments like Songdo (South Korea), Masdar (Abu Dhabi) or Sidewalk Toronto (Canada) are ultimately urban planning exceptions that are intended to make it possible to demonstrate sensory connection and prognostic data analysis away from the chaotic reality of a city, with data being gathered on the defining category of urban transformation (Halpern et al. 2013).

Yet the introduction of technological innovations for the purposes of energy supply, transport systems or communication services and the associated adaptation of cities also leads to unexpected side effects related to digital and automated processes (Townsend 2013, Kitchin/Dodge 2017). In addition to the promised advantages of convenience, safety and economic growth, the recurrence of urban problem areas and risks (emergence of new inequalities, security and criminality risks as well as pollution) can hardly be avoided (Kitchin/Dodge 2017). The resulting issues consist less in the use of technological innovations than in the failure to adequately consider urban diversity and to reconcile the benefits with the social, cultural and spatial characteristics of an urban context and hence to maintain a historical perspective (Hajer 2014, Picon 2015). In view of the increasing service and mobility provision, cities run the risk of letting the discourse around local conditions for a liveable city in the age of digital transformation fall victim to an overhasty implementation of technological solutions (Hajer 2014: 16).

In the specific case of CAM, it should be considered that local path dependencies in the form of mobility cultures, settlement structures, business connections and established planning concepts are an obstacle to the desired optimization effects. The complexity of these historical interdependencies tends to reproduce a "car system" (Urry 2004: 27) and hence a "car city" whose infrastructure improvements paradoxically merely delay problems or cause more rebound effects (Sonnberger/Gross 2018, Schneidewind/Scheck 2013). In order to change existing mobility and consumption practices in line with a sustainable transportation revolution (see Chaps. 3.1, 3.2), it is instead necessary to consider CAM in combination with social innovations. As technological artefacts always go hand in hand with a redesigned social order, social innovation, which can be understood as a social change process, is closely connected to technological artefacts (Braun-Thürmann 2005). On the one hand, this means that the deployment of CAVs cannot be understood as a simple solution to contemporary transport and planning problems, but must be critically contemplated in relation to the city's structure and society (Guerra 2016). On the other hand, it is essential that local planning strategies understand uncertainties as a natural part of the process and see unexpected developments as an opportunity for something new. While pilot projects on CAM should be tested both spatially and socially as soon as possible, they should be accompanied by a public discussion and collective learning process. In order to answer the question of what a long-term vision of urban coexistence with CAVs would look like, various directions of development must first be analysed and a discussion must be held around what is and is not desired.

4.7.2 RELEVANCE OF REFLEXIVE PLANNING APPROACHES IN LIGHT OF NEW MOBILITY TECHNOLOGIES

Given the uncertainties and unpredictable developments that generally go hand in hand with innovation and transformation processes, flexible, explorative and reflexive approaches in planning theory and practice are gaining attention (Balducci et al. 2011, Roorda et al. 2014, Freudendal/Kesselring 2016, Bertolini 2017, Hopkins/Schwanen 2018). For example, the understanding of "adaptive" or even "evolutionary" planning is based on an incremental trial of innovations and loose regulations in order to ensure urban resilience, i.e. the ability to regenerate and self-organize in a crisis, when faced with global challenges (Bertolini 2007, Rauws 2017). In light of "robust planning goals" (Bertolini 2017: 147), which despite uncertainties can be deemed desirable for various futures, the real-world potential of technological innovations must first be tested by means of exploratory acts. Variations and selection processes must be taken into consideration both in the conception phase and in later planning and development stages in order to learn from experiences and be able to make changes (Bertolini 2017: 156).

In the discourse around "reflexive governance", two main readings of the concept are differentiated (Voß et al. 2006). In the first instance, reflexivity is understood as the state of governance in a modern world, which is constantly being confronted with the unintended consequences of earlier actions, indeed with risks and limits (Voß et al. 2006, Schwarz 2014).[1] A second reading concerns new strategies, processes and institutions which are initiated by self-confrontation (Voß/ Kemp 2006). The solution and planning approaches of modernity, which are based on scientific certainty and ultimacy, are replaced by discourse in a "reflexive modernity" (Schwarz 2014: 209). According to Stirling (2006: 260), reflexive strategies rest on the plurality and the conditionalities that underlie both scientific findings and technological potentials. Taking account of long-term systemic consequences and developing alternative strategies therefore requires a variety of perspectives and an expanded knowledge base (Stirling 2006: 258). Principles like non-finality, experimentation, fault tolerance, risk intelligence and acting in uncertainty (Voß et al. 2006, Heidbrink 2007, Schwarz 2014) are taking on greater significance, as is the conception of long-term strategies that create the framework for short-term, exploratory actions and enable cross-project learning (Lissandrello/Grin 2011). Reflexive strategies are relevant for planning in the sense of both concomitant monitoring and process design, for example in the form of open dialogues to develop visions of the future or exploratory urban interventions. The alternative practices necessary for transformation pathways require not least reflection on the respective structural conditions (Grin 2006, Lissandrello/Grin 2011).

Yet it is not only within the planning discourse that urban interventions and experimental spaces have gained attention in recent decades (Heyen et al. 2018). Especially in the context of sustainability and transformation research, real-world trials have increasingly been encouraged for the purposes of applied research and analysing change processes (Schneidewind/Scheck 2013). In these trials, cities or city districts serve as geographically, temporally and institutionally delimited reference areas within which scientific and practical knowledge can be integrated by initiating and enabling social and technological change processes (Schneidewind/Scheck 2013: 240). Alongside the real-world laboratories frequently mentioned in the transformation debate in the German-speaking world, a range of international research approaches should be noted, including (sustainable or urban) living labs, (urban) transition labs and (sustainable) niche experiments (Schäpke et al. 2017).[2] As no standardized methodical practice has yet been established in the research community, real-world laboratory approaches in the German-speaking world have

varied in terms of individual characteristics like the understanding of transformative research, the carrying out of real-world experiments, transdisciplinary types of cooperation and further contributions to social change processes by means of sharing and conveying generated knowledge (Schäpke et al. 2017).

If urban real-world laboratories are understood as a transdisciplinary framework, they can serve to correlate various real-world experiments, to define a shared objective and to enable overarching knowledge sharing and reflection processes (Beecroft et al. 2018: 77). In accordance with transformative research, the research, practice and educational aims of a real-world laboratory, of the individual projects and of the involved actors must be coordinated. Even if collaboration on an equal footing is aspired to, it is difficult to avoid conflicts of interest and objective as well as epistemic confrontations (Dusseldorp 2017, Singer-Brodowski et al. 2018). Some real-world experiments can be categorized as being methodically between pure knowledge application and knowledge creation (Schneidewind/Scheck 2013: 241), with situation-specific settings in particular, similar to transition experiments, aiming to contextualize technological or ecological possibilities and to socially empower radical alternatives. Through continuous (self-) reflection and evaluation, an experimental-reflexive research style (Beecroft et al. 2018: 78) can contribute to processes being adapted to new or unexpected conditions and to the real-world laboratory as a whole being able to learn (Flander et al. 2014).

In light of technological innovations like CAVs, real-world laboratories constitute a relevant instrument for urban planning in terms of transformation research in order to launch a collective learning process with the involvement of civil society. Especially issues like changing mobility behaviours, the accessibility of and barriers to new mobility services and alterations to urban planning are suited to such laboratory situations. The opportunities for and limits to generating local knowledge or even to initiating change will be determined by the integration of various target dimensions and the prioritization of research principles. For urban development, it remains imperative to constantly ensure the social legitimacy of the objectives, the process design and the results (Schäpke et al. 2017).

4.7.3 INCLUDE CONNECTED AND AUTOMATED VEHICLES IN PLANNING DESPITE UNCERTAINTIES?

With regard to the formative phase of CAM and its urban deployment, reflexive, explorative and adaptive processes can be significant in several ways as part of strategic planning. First, they can counter a purely technology-driven rationality that strives to minimize the complexity and uncertainties and maximize risk control with the aid of fast, standardized solutions. By contrast, it is important to pursue planning approaches that focus on the conditions and necessary capacities so that urban development can take place under various future conditions (Rauws 2017). In the sense of an "adaptive" or "incremental" planning approach (Rauws 2017: 36), a strategic framework and an orientation aid can be created with development guidelines that preserve context-specific flexibility without stipulating specific spatial function designs or coalitions of actors in advance. Given the new landscape of actors and IT companies' increasing advances in knowledge, public competence development and knowledge creation are particularly necessary. In addition to conveying multisector interests, it is up to urban planning to accommodate the complexity of urban transformation processes and pursue measures that may be conducive to a variety of scenarios and urban lifestyles (Guerra/Morris 2018). Here it is less a matter of finding spatial or technological solutions than including spaces of possibility in urban planning that are open for future developments, i.e. strengthening the spatial and social ability to adapt and react.

Furthermore, the approaches, structures and systems should be contemplated that produced and preserve a car system in order to counteract its perpetuation with CAVs. To achieve this, proactive planning approaches are necessary through which contentions, confrontations and changes of perspective are voiced and made possible amongst planners, policymakers, scientists, businesspeople and civil society. In open dialogues and transdisciplinary projects whose aim is the long-term transformation of social practices, actors can be encouraged to discuss existing

challenges and a necessary adaptation in behaviour. With the aid of reflexive strategies, problematic situations can be brought to light, expectations of the future specified and structural constraints contemplated in light of existing routines (Lissandrello/Grin 2011). Such processes serve to encourage not merely communication between the actors, but also the integration of various perspectives in order to develop a shared view of reality, even if it includes various understandings of the problem, aims and strategies at first (Voß et al. 2006). The plurality of views and the collective creation of knowledge come to the fore of the collaboration (Voß/Kemp 2006). When calling attention to local development possibilities and synergies between the existing and the new, it is necessary to realign stakeholders' imaginations and identify alternative visions of the future for a city region or district. Ideally, previous understandings of urban mobility will change, while individual demands will approximate alternative future pathways (Lissandrello/Grin 2011: 244). In this way, reflexivity in planning, which also includes questioning the role and the underlying convictions of planners themselves (Beecroft et al. 2018: 92), can trigger a change process through productive confrontations (Lissandrello/Grin 2011: 226). The context for critical explorations of the future creates a framework for transformative practices that go beyond communicative and strategic planning principles.

Finally, the suitability of new mobility services for a specific spatial and social context can be checked or voted on using explorative planning approaches. To date, trial runs of highly automated vehicles (Level 4) have taken place in controlled orientation and experimentation contexts in

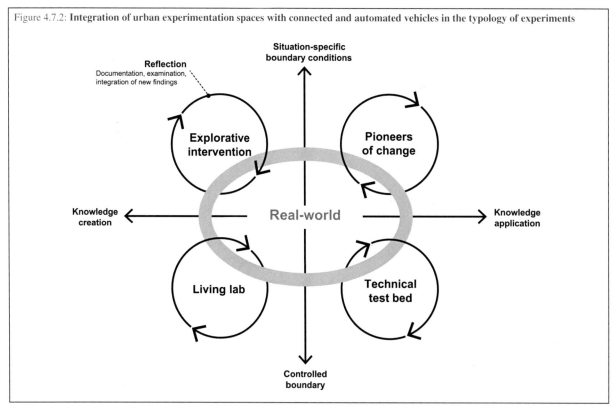

Figure 4.7.2: Integration of urban experimentation spaces with connected and automated vehicles in the typology of experiments

order to acquire transferrable knowledge for the purpose of technology development (see Chap. 4.1). Thus, the real-world trials of CAM can be characterized as being between controlled test beds, i.e. demonstration sites, and living labs, i.e. real-world tests with the involvement of users (see Fig. 4.7.2). Aspects like changes in mobility habits, interactions in the urban space and the diverse user needs of target groups like people with limited mobility have hardly been taken into consideration in connection with CAM. From the point of view of the city and urban planning, it must therefore be established what insights can be gained from the studies conducted thus far and to what extent future trials can be methodically expanded and co-created.

Moreover, explorative trial actions in the form of short-term urban interventions or real-world urban laboratories can help to answer transformation-related questions in an exemplary way and gain location-specific experience. However, the requirement should be to pursue aims that are socially legitimized, ethically well founded and for the common good (Defila/Di Giulio 2018). From a scientific perspective, specific knowledge of the system, objective, transformation and process should be surveyed locally, the social relevance of the respective research questions should be tested by involving practice-oriented actors and the focus of individual real-world experiments should be chosen collaboratively (Beecroft et al. 2018: 80).

In addition, those local actors should be taken into consideration who concern themselves with the impact of new mobility services on public space, working conditions, communities, the climate, education, accessibility and affordability. Triggering alternative mobility practices and different means of use would make it possible to acquire practical knowledge about unintended consequences, the transferability of results and the effects and requirements of forms of cooperation and empowerment (Beecroft et al. 2018: 81). If issues and transformation aims are locally grounded and self-organized groups are empowered to propose their own ideas for solving problems, then initiatives can arise that generate cultural, social and spatial added value and bring about cultural change as "change agents" or pioneers of change (WBGU 2011: 256). This would require not only an openness to sociocultural particularities and collaboration on equal terms, but also the possibility for local actors to participate beyond the time frame of mere project stages. Systematic reflection on the collaboration and the results can ultimately serve various education processes, such as the ability to learn and adapt within real-world laboratories or the formulation of transformation pathways as social learning processes in the broader sense (Schneidewind/Singer-Brodowski 2015). Creatively reconfiguring the present and experimenting with the ability to change make reflexive and explorative processes valuable niches in which the limits of what is possible are retested and urban diversity as a quality is maintained (Abbott 2012).

1   This refers to "reflexive modernity" according to Beck (1986) and others (Beck et al. 2003), meaning the fundamental transformation of a modern society and its political systems. In contrast to the views of a postmodern era, with the concept of reflexive modernity the authors assume a restructuring and reconceptualization of social conditions and historical ruptures. According to Latour (2003), reflexivity does not mean the increase in controllability and consciousness, but rather the awareness that complete controllability is effectively impossible.

2   According to Schäpke et al. (2017: 49), the three approaches should be differentiated with regard to methodical technique, focus and underlying theories: (1) living labs aim at marketable, standardized products and services as well as generalizable findings – and hence they limit participation and strive for controlled experiments and settings; (2) transition experiments target evolutionary forces, the empowerment of pioneers and reinforcement of alternatives – as well as the orchestration of experiments and specific, real-world change; consequently, the approach focuses on strong forms of participation and empowering mechanisms of disseminating alternatives through learning; (3) niche experiments contribute to the generation of reflexive knowledge through their eschewal of conducting their own experiments; depending on the projects and processes analysed, this is highly contextualized or more generalizable.

SCENARIOS

SHAPING CHANGE AT THE LOCAL LEVEL DURING THE TRANSITION PERIOD

5

5.1

SCENARIO DEVELOPMENT AND STRUCTURE

Developing scenarios means producing multilayered visualizations of a possible future that take into account economic, technological, social, policy and planning aspects. Starting with an analysis of the current situation, they demonstrate – quantitively or normatively and narratively (Kosow et al. 2008: 52–55) – often idealized options, which are normally characterized by different interests, objectives and interventions (Schulz-Montag/Müller-Stoffels 2006, Wilms 2006, Heinecke 2012, Fagnant/Kockelman 2014b).

In principle, planning is based on a compromise between desired aims and existing means, and factors in the uncertainty of unanticipated consequences, rebound or lock-in effects. In the context of this project, normative and narrative scenarios were thus developed for application in urban and transport planning, for public administrations, policy and urban and mobility research. The approach included the unique characteristics and current challenges of the European city (see Chap. 3.2). The narrative scenarios method aims to point the way out of regulatory, largely short-term thinking in (supra-)local spatial and transport planning, instead encouraging long-term alternative thinking and offering new perspectives and ways of looking at problems (Minx/Böhlke 2006, Kosow/León 2015).

Due to the objective of developing scenarios that would be relevant for steering decisions, attitudes to policy and planning were declared a key factor. The scenario narratives that

built on this were elaborated on in focus groups and then checked for consistency and/or plausibility. As a communication aid, graphics and illustrated scenes were used which were repeatedly revisited and improved throughout the entire process.

Formulated in writing and then illustrated, the scenarios are an important communication tool in a collaborative planning process when it comes to developing conceivable futures, formulating concrete aims for the introduction of CAVs and identifying risks as well as jointly seeking and negotiating desirable solutions in the areas of mobility, settlement and neighbourhood development. The scenarios also make it clear that the future is malleable – at least within defined limits. The questions of whether and how it can be shaped depend on local constellations and the actors involved as well as their interests. In addition to the many prognostic scenarios now available, we take as our starting point the fundamental malleability of the technological transformation process in cities and regions.

Figure 5.1.1: **Overview of the scenarios' structure**

THE THREE SCENARIOS DIFFER IN TERMS OF THE …	THE SCENARIOS ARE PRESENTED ON SEVERAL LEVELS:
■ … attitude of policymakers and planners underlying the decision-making processes,	■ Description revolving around the attitude of policymakers and planners
■ … groups of actors involved and the creative scope they were given,	■ Normative narratives with the support of comics to illustrate the scenarios
■ … weighting of key challenges for urban development and the means used,	■ Tabular comparison of the scenarios' main parameters, elements and factors
■ … locations treated as a development priority,	■ Stakeholder assessment of the potentials of the various attitudes to policy and planning
■ … the applied CAV use cases as well as their fields of application and integration in the existing mobility system.	■ Reflection on and generalization of possible socio-spatial impacts

5.2

KEY FACTOR: WAYS TO STEER POLICY AND PLANNING

In current discussions about the tasks and forms of policy, there is often talk of a change in statehood – from government to governance. The term "governance" is used for various means of regulating collective issues which differ in terms of the type of self-management and the way they address power (Hamedinger 2013: 56–57). The political concept of governance has gained in significance as a form of government and management since the 1990s (Heeg/Rosol 2007: 504). Starting in the 1980s, entrepreneurial strategies have been introduced in public administration with increasing frequency (Heeg/Rosol 2007: 497), with administrations relying more and more on private sector management techniques (New Public Management). Although this has not entirely replaced the classic form of governance, it has been expanded and transformed by new actors from the private sector and civil society. Over the course of this period, public sector responsibilities have also changed, primarily as a result of the privatization of public companies, the outsourcing of public services and the commercialization of what remains of the public sector (Jessop 2002).

The change towards this form of governance means that the decision-making processes of policymakers and planners now involve (transnational) companies, NGOs and civil society alongside the formal, legitimized state actors and institutions (Heeg/Rosol 2007: 504; Hamedinger 2013: 62). This occurs on the one hand due to "outside pressure" and increased interest in participation, and on the other the political administrative system is dependent on economic, technical and social innovations. As a result, processes and decisions become more flexible, reversible and generally quicker. Consequently, not only do the actors involved change, but so too do their duties, responsibilities, competences and ability to exercise power.

The discussion below explores three different forms and interests of steering policy and planning which constitute the foundation for conceptualizing the scenarios. The focus here is on the three sectors market, state and civil society, all of which have acquired new significance as governance has become established.

5.2.1 THE ATTITUDE OF THE MARKET

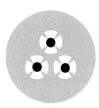

With the concept of governance, a policy of flexibilization and decentralization is pursued that is mostly accompanied by a weakening of state-centred influence on leadership, power and steering (Ansell/Torfing 2016: 2). This approach supports the deregulation of the public sector and the belief that private and (semi-)public markets can unleash self-regulating forces which are more flexible, rapid and efficient. The role of elected governments is limited to their accountability for general political objectives, while the responsibility for the production or operation of public services is put in the hands of private contractors or publicly appointed, specialized agencies that are driven entirely by economic principles (Ansell/Torfing 2016: 6). Competition thus becomes more positive because citizens can choose between a wide range of different providers on the free market.

This understanding of steering policy and planning is at the heart of the market-driven approach. When (the very heterogeneous) market interests are put at the centre of steering processes, quite neoliberal objectives are pursued, companies' influence increases and the triple bottom line shifts in favour of financial competitiveness. For the introduction of CAVs, that means that tech-

Figure 5.2.1: **Summary of the attitudes underlying the scenarios**

	MARKET-DRIVEN SCENARIO	POLICY-DRIVEN SCENARIO	CIVIL SOCIETY-DRIVEN SCENARIO
Diagnosis	Key challenges are also opportunities and can be solved using the right technologies.	Key challenges can be overcome with the right tools by the responsible level of administration.	Key challenges can be tackled at local level and with people's common knowledge.
Inherent logic	Efficiency, competitiveness	Legitimized by the public	Empowerment of local groups of actors
Assumption (target system)	Mobility as a business model	Multimodal mobility system	Sufficiency, appropriation of technologies
Prerequisites	Dismantling barriers to the private sector	Public acceptance	Change in power relations, change in values

nological progress is considered in terms of economic interests. With regard to connection and automation, the main focus is on the potential of new business models, diverse use cases and new mobility providers between the conflicting priorities of international economic interests (Google, Amazon, Uber, Lyft etc.) and local/regional steering of policies and planning.

5.2.2 THE ATTITUDE OF POLICY

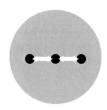 Following the classic understanding of steering as government, elected political decision makers assume responsibility for policy and planning decisions – on the basis of transparent departmental planning – but increasingly involve stakeholders. The (municipal) government uses state and municipal resources in the interests of the common good, and draws up and implements political programmes. Cities and public institutions are "paternalistic distribution agencies" (Heeg/Rosol 2007: 493) that not only distribute public funds but also assume responsibility at administrative level for citizens' welfare. In this context, there is great confidence in the public welfare system, supported by the assumption that all citizens' needs are treated as equally as possible.

The self-conception of policy and planning has adapted in order for a well-functioning state to remain a prerequisite for successful spatial, economic and social development. This becomes apparent in the fact that new corporatist forms of negotiation have been introduced to pursue local and regional development goals that are increasingly aligned with the interests of sustainable development in the city region. The aims, content, processes and procedures of the political representation of interests are now defined proactively. Despite initiating negotiation processes between public and private market actors, the public sector strongly regulates the activities and reach of the private sector. The alliances with private companies, also in transport planning, are very much structured in favour of the political aims of ecological and social sustainability. This understanding of governance is at the heart of steering processes in the policy-driven approach.

5.2.3 THE ATTITUDE OF CIVIL SOCIETY

 From the perspective of social groups and communities, governance is largely understood as self-organized processes by civil society. Compared to being steered by the state or market, governance by civil society is considered more consensus-oriented, egalitarian, trust-based and deliberative because the intrinsic values of civil society can be better met and mobilized. Accordingly, governance becomes the reflexive self-organization of independent actors (Jessop 2003: 1), with civil society-based forms of regulation being considered an important counterweight to institutionalized, bureaucratized state structures.

In this approach, important parts of local and regional sub-state power are transferred to the local level, with steering by civil society being considered important due to its intrinsic knowledge, mutual trust and community solidarity. Social communities thus represent a "bottom-up" organization with the potential for grassroots democratic reform of the state sector. Since the 1980s, some social movements have developed into professional NGOs and become increasingly politically relevant (Brand et al. 2001). In many contexts, participation and cooperation are seen as necessary conditions for effective government.

Civil society is also considered a resource for social innovation beyond political steering, either serving the group's purpose directly or being used for economic and/or policy and planning processes (Dangschat 2017b). Consequently, alongside policy, market actors also attempt to set in motion civil society processes for economic purposes.

Civil society has the potential to provide numerous impulses for socio-spatial change processes, a circumstance that is repeatedly emphasized in sustainability debates. This understanding of governance is at the heart of steering processes in the civil society-driven approach.

5.2.4 CAT-DRIVEN TRANSFORMATION UNDER THE THREE APPROACHES

In recent decades, the notion of transformation has become one of the most important paradigms – not only in the scholarly discussion on sustainability, but also in political agendas (Koch et al. 2017: 1). In light of the current development goals for the European city, this transformation should be understood as developments towards inclusivity and sustainability (UNECE 2012). A major role is accorded to cities, especially when it comes to fundamental and multiple changes (WBGU 2016; Koch et al. 2017: 1). Although the term "urban transformation" is widespread, a clear and consistent definition is still lacking (Koch et al. 2017: 1). Stirling (2014) differentiates between "societal transitions" and "social transformations".

Societal transitions include technological innovations with which predefined goals can be achieved. Social transformations, meanwhile, are the result of diverse, slowly emerging, fractious political realignments that encompass both social and political but also technical aspects and do not pursue any clearly predefined goal (Koch et al. 2017: 3).

Precisely this openness to possible socio-technical developments – as a result of digitalization in general and the introduction of CAVs in particular – is illustrated by the focus of the three outlined scenarios. The different focal points are challenged and discussed with reference to the technical developments associated with the introduction of CAVs. Just how the balance of power will shift remains open, but it is clear that it will have a profound influence on the future implementation of new technologies and their acceptance. The form this might take in each of the scenarios is modelled in Figure 5.2.2.

Figure 5.2.2: **Forms of urban governance: actors, structures and processes**

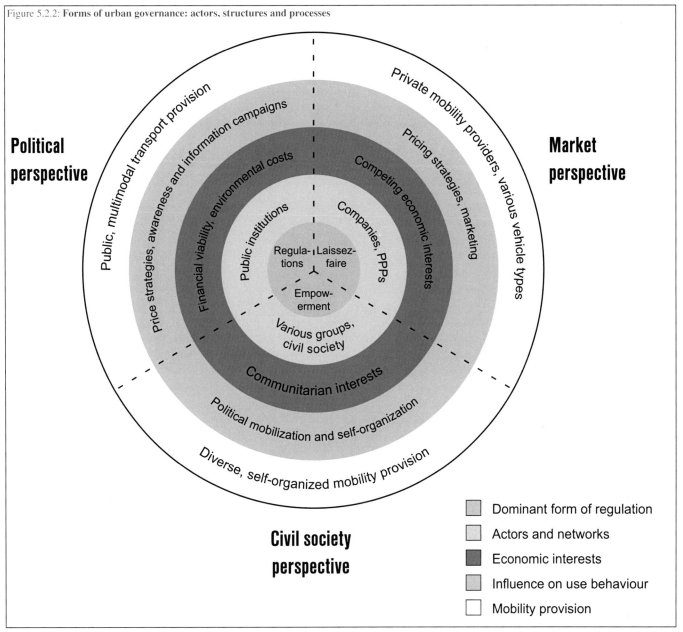

Political perspective

Market perspective

Civil society perspective

Public, multimodal transport provision

Price strategies, awareness and information campaigns

Financial viability, environmental costs

Public institutions

Regula-tions

Laissez-faire

Empow-erment

Companies, PPPs

Competing economic interests

Pricing strategies, marketing

Private mobility providers, various vehicle types

Various groups, civil society

Communitarian interests

Political mobilization and self-organization

Diverse, self-organized mobility provision

Dominant form of regulation

Actors and networks

Economic interests

Influence on use behaviour

Mobility provision

Source: AVENUE21 based on WBGU (2016: 107)

5.3 MARKET-DRIVEN APPROACH

MOBILITY
Customized mobility packages

CITY REGION
Locations with optimized functions

NEIGHBOURHOOD
IoT neighbourhood

The market-driven scenario is dominated by customized service packages offered by private providers, which are experienced as part of automated "ecosystems". With the introduction of CAT, increased efficiency is achieved and premium customers are prioritized on motorways and specially upgraded routes in the city region. In the higher price segment, people mostly travel alone; lower occupation rates are the rule. In addition to customized premium offers, there are also sharing services that are cheaper but less comfortable. Passengers' attentions and journey times are monetized thanks to data-based business models. The city region is similarly characterized by contrasts: residential, industrial and business parks or "edge cities" with optimized functions have emerged along the motorways, thus connecting them with an international, CAV-compatible, large-capacity road network. In areas that cannot be reached by CAVs, retail parks and production sites come under extreme pressure.

Political and administrative actions are determined by a restrictive fiscal policy. Urban process steering and infrastructure development (CAV-compatible routes, pick-up/drop-off zones and sensors in the streetscape) are outsourced to private companies (competitive outsourcing) or developed by PPPs. In line with the idea of the global city, the city feels obliged to compete with other towns and cities around the world (for companies and workers). Communication and cooperation with the region beyond the city limits exist only where necessary. Favourable conditions for innovation and market-oriented funding programmes are intended to produce solutions to urban challenges. Hopes are pinned on the IoT and Industry 4.0 in this scenario. Market

deregulation and the expansion of test areas and statutory experimentation clauses in "smart" neighbourhoods are supposed to accelerate economic development and technological progress. Unlocking hidden potential, increasing efficiency, consolidating resources and optimizing the system are all considered progressive.

Progressiveness and innovativeness are seen as virtues. Self-optimizers enjoy wide social acceptance. Society largely lives on digital networks, is internationally mobile and enjoys the convenience of e-commerce's globalized market diversity. Individualization, flexibilization and pluralization processes develop, shaping the world of work and everyday life. It is possible to go to both the library and the pharmacy late at night and not just during the day. In such places, customers are served not by people but by accommodating mobile bots who – regardless of the time of day or night – are always extremely patient. Individuals' demand for autonomy and flexibility leads to enormous land use and high infrastructure costs in the rapidly growing metropolitan region. It is now hardly possible to curb suburbanization or reduce the increasing swarms of commuters; transport emissions repeatedly exceed the maximum limits. In many cases, public space has been privatized. Although the quality of the design and time spent in such spaces is high, their users must have purchasing power.

"By integrating automated vehicles into our mobility platform, we can offer our customers even more individuality and freedom.

Our range comprises various mobility packages that enable a seamless, secure and convenient travel experience. Users select their comfort category and destination via the app and can then choose from a range of offers that include journey time, routes and costs. Users can book their journey in advance or on demand and can see in real time when their vehicle will arrive at the pick-up point.

Via the shuttle's integrated infotainment system, additional offers like entertainment media, stops en route or instantaneous route changes can be selected during the journey. It is also possible to pay extra for premium services like priority routes during rush hour. Travelling via side roads, adding longer walking distances to the pick-up point and destination, as well as bookings outside of rush hour enable a tailored service for people with lower incomes.

Individual mobility and comfort preferences are saved on the user's profile and immediately taken into account for their next booking."

"NAVIGATING THE VEHICLES IN OUR FLEET IS BASED ON THE INTEGRATION OF VARIOUS REAL-TIME DATA. AS A RESULT OF ITS CONNECTION AND COMMUNICATION WITH OTHER VEHICLES AS WELL AS ENVIRONMENTAL SENSORS, THE NAVIGATION SYSTEM CAN REACT TO AND EVEN PREDICT THE UPCOMING TRAFFIC VOLUME AND THEN DYNAMICALLY ADAPT ITS ROUTE, FOR EXAMPLE. WE ARE WORKING ON INTEGRATING DATA FROM MOBILE TERMINALS OR INTELLIGENT WEARABLES INTO OUR ALGORITHMS IN THE FUTURE. BY INCLUDING OTHER INTERFACES, WE AIM TO IMPROVE OUR PROGNOSTICS. ON THE BASIS OF PERSONALIZED DATA, ASSISTANCE SYSTEMS SHOULD BE ABLE TO RECOGNIZE THE REASON FOR A JOURNEY AND ANTICIPATE INDIVIDUAL NEEDS OR SPONTANEOUSLY PICK UP PARCELS AND PURCHASES.

THIS WILL MAKE MANAGING THE FLOW OF USERS DYNAMIC, SMOOTH AND EFFICIENT. THE LONG-TERM UPGRADE OF OUR FLEET TO FULLY AUTOMATED VEHICLES WILL FURTHER INCREASE THE TEMPORAL AND SPATIAL CONVENIENCE ADVANTAGES OF OUR SERVICE. WHAT ACTIVITY IS COMPLETED AT WHAT TIME IN WHAT PLACE WILL TAKE A BACK SEAT IN FUTURE DUE TO THE HIGH LEVEL OF FLEXIBILITY. WITH OUR MOBILITY SERVICE, WE WANT TO SUPPORT THE ADVANTAGES OF SUCH FREEDOMS."

"IN ORDER TO MAKE THE OVERSTRETCHED TRANSPORT NETWORK MORE EFFICIENT, THE CITY REGION IMPLEMENTED A TRANSPORT CONTROL ROOM THAT AUTOMATICALLY ANALYSES AND STEERS TRANSPORT FLOWS. WE BENEFIT FROM THAT AS A MOBILITY PROVIDER BECAUSE WE RECEIVE PRECISE TRAFFIC INFORMATION FROM THE CITY REGION AND INCORPORATE IT IN OUR FLEET'S ROUTING ALGORITHMS. YET DESPITE THE INCREASED EFFICIENCY OF THE FLOW OF TRAFFIC, THE COMMUTER TRAFFIC FROM THE SUBURBS IS STILL ENORMOUS. WHICH IS WHY THE MUNICIPAL GOVERNMENT DECIDED TO GRADUALLY INTRODUCE DYNAMIC PRICING FOR THE MOBILITY NETWORK.

THIS IS INTENDED TO REDUCE THE BUILD-UP OF TRAFFIC JAMS ON ACCESS ROADS AND COVER THE NEED FOR INFRASTRUCTURE INVESTMENTS. AS A MOBILITY SERVICE PROVIDER, WE ARE INTEGRATING THESE COSTS INTO OUR PRICES AND ADAPTING THEM ACCORDING TO DEMAND AND DESIRED OCCUPANCY LEVEL."

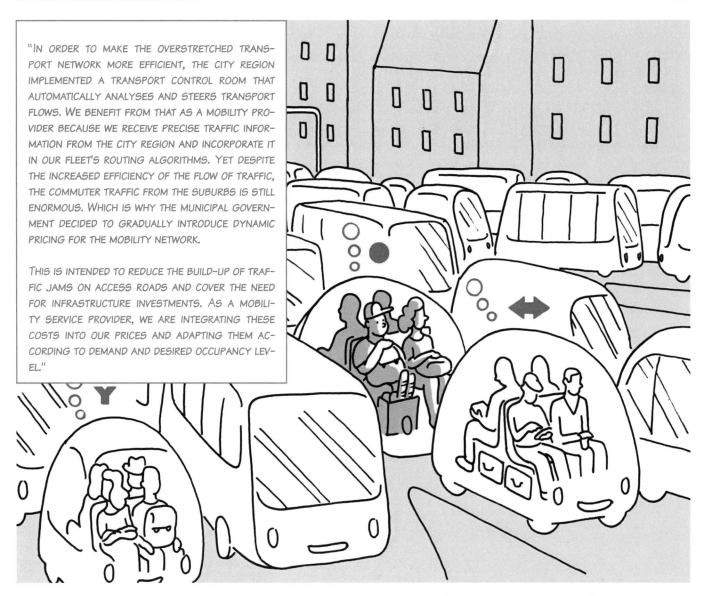

"AS WE SHARE CERTAIN TRAFFIC DATA COLLECTED BY OUR FLEET WITH THE CITY, WE ARE ABLE TO NEGOTIATE RIGHTS OF WAY AND DISCOUNTS. SINCE THE DYNAMIC PRICING SYSTEM WAS INTRODUCED, OUR CLIENTELE ON THE OUTSKIRTS HAS BEEN GROWING STEADILY AS A RESULT OF OUR SHARING OFFERS.

CONSEQUENTLY, WE NOW PROFIT FROM RESIDENTS MOVING TO THE OUTSKIRTS DUE TO THE RISING LIVING COSTS IN THE INNER CITY, BECAUSE THE HIGHER THE USER DENSITY IN AN AREA, THE MORE VEHICLES WE SEND THERE. AT THE MOMENT, WE ARE COLLECTING THE NECESSARY DATA TO BE ABLE TO STEER THE FLEET DISTRIBUTION COST-EFFICIENTLY DEPENDING ON THE TIME OF DAY ONCE AUTOMATED VEHICLES ARE INTRODUCED ACROSS THE ENTIRE CITY."

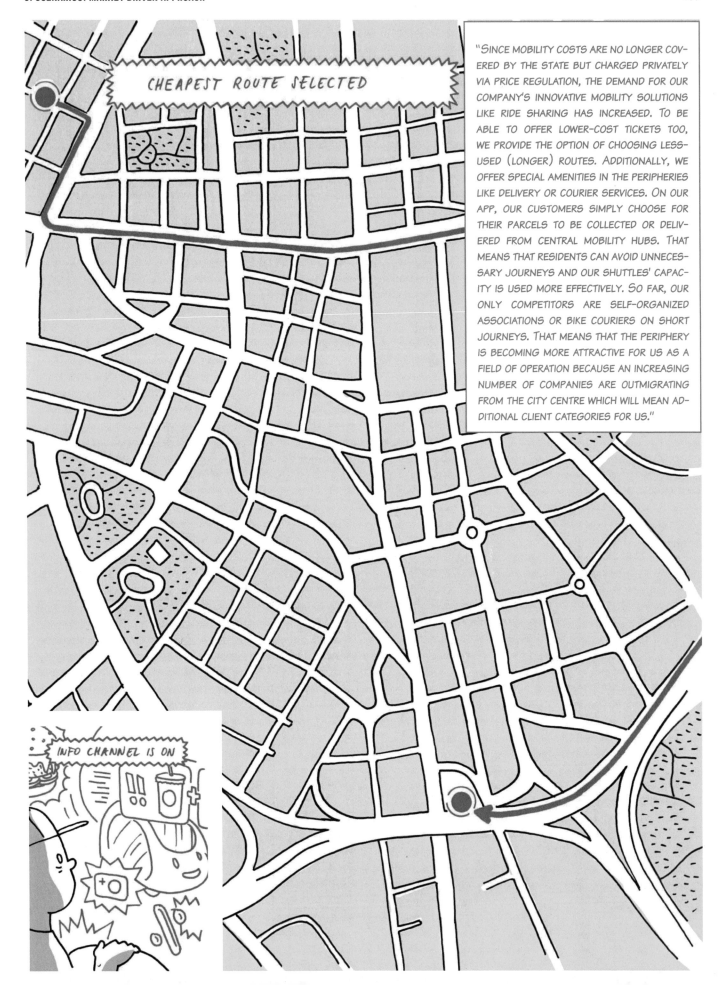

CHEAPEST ROUTE SELECTED

INFO CHANNEL IS ON

"SINCE MOBILITY COSTS ARE NO LONGER COVERED BY THE STATE BUT CHARGED PRIVATELY VIA PRICE REGULATION, THE DEMAND FOR OUR COMPANY'S INNOVATIVE MOBILITY SOLUTIONS LIKE RIDE SHARING HAS INCREASED. TO BE ABLE TO OFFER LOWER-COST TICKETS TOO, WE PROVIDE THE OPTION OF CHOOSING LESS-USED (LONGER) ROUTES. ADDITIONALLY, WE OFFER SPECIAL AMENITIES IN THE PERIPHERIES LIKE DELIVERY OR COURIER SERVICES. ON OUR APP, OUR CUSTOMERS SIMPLY CHOOSE FOR THEIR PARCELS TO BE COLLECTED OR DELIVERED FROM CENTRAL MOBILITY HUBS. THAT MEANS THAT RESIDENTS CAN AVOID UNNECESSARY JOURNEYS AND OUR SHUTTLES' CAPACITY IS USED MORE EFFECTIVELY. SO FAR, OUR ONLY COMPETITORS ARE SELF-ORGANIZED ASSOCIATIONS OR BIKE COURIERS ON SHORT JOURNEYS. THAT MEANS THAT THE PERIPHERY IS BECOMING MORE ATTRACTIVE FOR US AS A FIELD OF OPERATION BECAUSE AN INCREASING NUMBER OF COMPANIES ARE OUTMIGRATING FROM THE CITY CENTRE WHICH WILL MEAN ADDITIONAL CLIENT CATEGORIES FOR US."

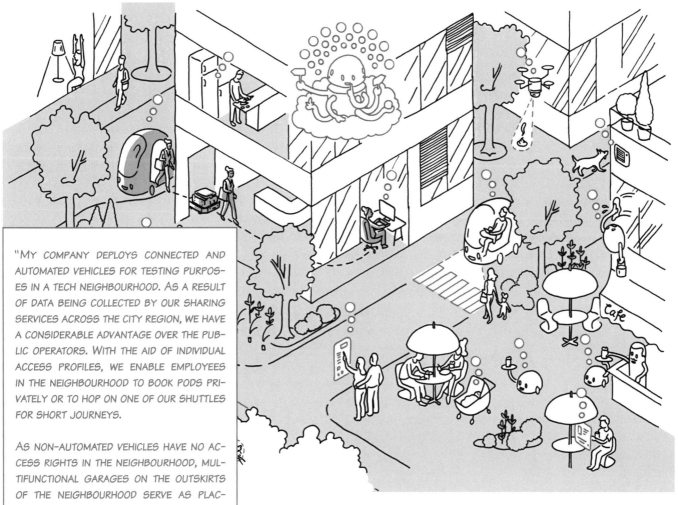

"MY COMPANY DEPLOYS CONNECTED AND AUTOMATED VEHICLES FOR TESTING PURPOSES IN A TECH NEIGHBOURHOOD. AS A RESULT OF DATA BEING COLLECTED BY OUR SHARING SERVICES ACROSS THE CITY REGION, WE HAVE A CONSIDERABLE ADVANTAGE OVER THE PUBLIC OPERATORS. WITH THE AID OF INDIVIDUAL ACCESS PROFILES, WE ENABLE EMPLOYEES IN THE NEIGHBOURHOOD TO BOOK PODS PRIVATELY OR TO HOP ON ONE OF OUR SHUTTLES FOR SHORT JOURNEYS.

AS NON-AUTOMATED VEHICLES HAVE NO ACCESS RIGHTS IN THE NEIGHBOURHOOD, MULTIFUNCTIONAL GARAGES ON THE OUTSKIRTS OF THE NEIGHBOURHOOD SERVE AS PLACES TO CHANGE TRANSPORT MODE OR FOR TRANS-SHIPMENT. THAT MEANS THAT PUBLIC SPACE IS LARGELY CAR-FREE. BETWEEN THE LANE MARKERS FOR THE SELF-DRIVING VEHICLES AND THE PAVEMENT, THERE IS A FLEX ZONE WHOSE USE ADAPTS TO DEMAND OVER THE COURSE OF THE DAY.

SERVING AS A LOADING AREA FIRST THING, BY MID-MORNING IT IS USED AS AN EXTENDED PAVEMENT. THEN AT THE WEEKEND, IT IS TRANSFORMED INTO A KERBSIDE SEATING AREA FOR THE NEARBY CAFÉS AND RESTAURANTS. CONSEQUENTLY, THE STREETSCAPE IS STILL REGULATED TO PRESERVE A SENSE OF SECURITY. THIS ALSO MAKES IT EASIER FOR ENVIRONMENTAL SENSORS TO COMMUNICATE THE USE INFORMATION AND MAINTENANCE NEEDS TO THE NEIGHBOURHOOD CONTROL ROOM."

"As someone who loves technology, I am convinced that in the long term every aspect of life will be integrated into an intelligent ecosystem. In the future, my personalized virtual assistant would automatically inform me about the weather and recommend the most appropriate means of transport and route. An automated shuttle would soon be available because my routine would have been saved. During the journey, my calendar would pop up and remind me of upcoming appointments or unanswered messages.

Any food I need would be delivered in my shuttle every evening. To achieve that, goods and passenger transport would have to be better integrated, not just virtually but also in the urban space. The retail areas in the centre of the neighbourhood would be reduced to the absolute minimum but connected with my user profile via digital interfaces. It would not only be possible to view special offers, traffic information or current patient numbers at the health centre in real time and specific to my location, but the technology would predict my decisions and make my everyday life easier."

5.4 POLICY-DRIVEN APPROACH

MOBILITY
Multimodal public
mobility systems

CITY REGION
Cooperation with
the city region

NEIGHBOURHOOD
Multimodal traffic hubs
as neighbourhood centres

In the policy-driven scenario, an integrated multimodal transport network forms the backbone of mobility and settlement development. To the greatest possible extent, the physical and digital infrastructure in the city region is built and operated publicly, also in order to maintain sovereignty over the data it generates. A publicly operated, integrated mobility platform simplifies intermodal travel with public transport (buses, trams and underground trains), automated shuttles, car- and ride-sharing services and e-bikes or scooters. Via the platform and the data it generates, means of transport are purposefully prioritized (according to their environmental impact) or incentives are offered to use active mobility. Automation and digitalization are intended to serve the public good. There is intensive cooperation in the city region regarding matters of settlement, infrastructure and mobility development. In order to counteract the increase in traffic caused by delivery services, stretches of land in the region are set aside, the delivery network in the city region is strategically planned and distribution centres are developed.

The movement of people and goods is concentrated in a hierarchical system of mobility hubs that act as intermediaries between areas with higher and lower CAT compatibility and make it possible to change mode of transport conveniently and safely in line with the notion of "seamless transport".

Everyday functions and public institutions are likewise planned around these hubs in order to reduce journeys. New urban centres arise in these hubs that stimulate neighbourhood development. In the context of such transit-oriented developments (TOD), the motorway network is also integrated into the system, with transfer zones being purposefully developed where passengers can change between (supra-)regional and local mobility. Within neighbourhoods, high urban density and mixed use are encouraged in order to increase suitability for soft mobility. Urban sprawl is mostly prevented thanks to a restrictive land policy. In already developed areas between transport routes, connected and automated shuttles increase accessibility by public transport.

However, the public sector faces challenges and problems regarding financing in the city region: connected and automated mobility leads to a loss in income (e.g. parking tickets and fines, speeding tickets, etc.) and ultimately there is increased competition between public transport and private mobility providers, primarily in areas with high user frequency that are also CAT compatible. Use of the road and rail network is priced on the basis of existing and generated mobility data (e.g. mobility pricing or granting licences).

"IT WAS CLEAR THAT IN FUTURE SHAPING MOBILITY SERVICES AND SOLVING TRAFFIC PROBLEMS COULD ONLY BE ACHIEVED BY A STRONG PUBLIC SECTOR. WE SIMPLY HAVE YEARS OF EXPERIENCE IN THE PROVISION OF MOBILITY SERVICES. THE ADDITIONAL DEPLOYMENT OF CONNECTED AND AUTOMATED VEHICLES HELPS US TO OPTIMIZE PUBLIC TRANSPORT AND MAKE IT MORE FLEXIBLE IN LINE WITH OUR POLICY OF PROVIDING BASIC SERVICES, HENCE COUNTERACTING THE STRONG DEPENDENCE ON PRIVATE CARS — EVEN THOUGH FINANCING SUCH SERVICES WILL ALWAYS BE A CAUSE FOR CONTROVERSY IN SOME QUARTERS."

"WE WERE LUCKY THAT PEOPLE WERE ALREADY FAMILIAR WITH THE IDEA OF GETTING ON A BUS THAT THEY WERE NOT DRIVING THEMSELVES. THAT OBVIOUSLY HELPED US TO QUICKLY ATTRACT NEW CUSTOMERS TO OUR MOBILITY SERVICE. RIGHT NOW, IT HAS TO BE SAID THAT IT WAS GOOD THAT WE PUT COMPREHENSIVE MEASURES IN PLACE SO EARLY TO INTEGRATE AUTOMATED VEHICLES INTO OUR RANGE OF SERVICES: WE CAN STILL MANAGE PASSENGER AND FREIGHT TRAFFIC IN OUR CITY AND STEER THE NATURE OF THE TRANSPORT OFFERED."

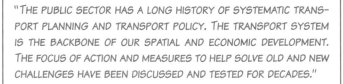

"THE PUBLIC SECTOR HAS A LONG HISTORY OF SYSTEMATIC TRANS-PORT PLANNING AND TRANSPORT POLICY. THE TRANSPORT SYSTEM IS THE BACKBONE OF OUR SPATIAL AND ECONOMIC DEVELOPMENT. THE FOCUS OF ACTION AND MEASURES TO HELP SOLVE OLD AND NEW CHALLENGES HAVE BEEN DISCUSSED AND TESTED FOR DECADES."

"WE WERE AWARE OF THE DISADVANTAGES OF PUBLIC TRANSPORT IN THE PAST, WHICH IS WHY IT WAS IMPORTANT TO MAKE CHANGING FROM ONE MODE OF TRANSPORT TO ANOTHER – BOTH DIGITALLY AND ARCHITECTURALLY – AS EASY AS POSSIBLE. ABOVE ALL, WE WERE HELPED BY THE EXTENSIVE COOPERATION THROUGHOUT THE CITY. NEVERTHELESS, IT REMAINS IMPORTANT TO MAKE OUR SERVICES MORE FLEXIBLE BY USING DIFFERENT TYPES OF VEHICLE AS WELL AS COMBINING AND REDUCING JOURNEYS IN PASSENGER AND GOODS TRAFFIC (E.G. LAST-MILE DELIVERY OF PARCELS) AS MUCH AS POSSIBLE."

"Since connected and automated vehicles were first launched and in light of the associated bad news regarding increased traffic volumes and suburbanization, we luckily realized that the development of our transport network and settlement had to be integrated and required collaboration at every level.

For that reason, we purposefully expanded the existing and tried-and-tested instruments in the area of settlement development (like polycentric development and strategic promotion of development along public transport routes) by also promoting and implementing connected and automated transport."

"Regulations coordinated at the city-regional level helped us to target settlement and transport development along large-capacity public transport routes in particular. This comprised not only optimizing existing railway lines, but also installing automated passenger transport on motorways with the same function as suburban railway lines, with stops at motorway exits. Furthermore, automation also helped us to provide an attractive public mobility service in areas between transport routes, i.e. in areas characterized by high car dependency, and to connect them with large-capacity public transport routes.

We are convinced that the transport system can be shaped by policy. Which is why we will continue to take responsibility for planning, organizing and operating our transport system in future."

"THE STRENGTHENED COOPERATION AT CITY-REGIONAL LEVEL REGARDING TRANSPORT AND SETTLEMENT DEVELOPMENT ALSO HELPED US TO BETTER COORDINATE WITH ONE ANOTHER WHEN DEVELOPING NEW NEIGHBOURHOODS. SOME AREAS AROUND REGIONAL AND SUBURBAN RAILWAY STATIONS WERE COMPLETELY REDEVELOPED TO SUPPORT MIXED FUNCTION AND MIXED USE.

THE RAILWAY STATIONS WERE EXPANDED INTO HUBS WITH NUMEROUS USES AND NOW FORM THE CENTRE OF THESE NEIGHBOURHOODS, WHICH ARE CHARACTERIZED ABOVE ALL BY PEDESTRIAN-FRIENDLY STREETSCAPES AND IMPROVED PUBLIC SPACE. THE GROUND FLOORS OF BUILDINGS IN THESE AREAS WERE ALSO VITALIZED TO FURTHER IMPROVE PUBLIC SPACE. THE CONCENTRATION OF TRANSPORT ROUTES, BOTH IN PASSENGER AND GOODS TRANSPORT, IS PARTICULARLY SIGNIFICANT.

THE USE OF AUTOMATED VEHICLES, MOSTLY IN THE FORM OF SHUTTLES AND PODS, HELPS PASSENGERS REACH THE HUBS FROM FURTHER AFIELD IN ORDER TO THEN CHANGE TO LARGER-CAPACITY MEANS OF TRANSPORT TO GET TO THE CITY CENTRE. TO THE GREATEST POSSIBLE EXTENT, THE HUBS HAVE BEEN BUILT TO MAKE IT EASY TO CHANGE MODE OF TRANSPORT. THAT MEANS THAT THE INHABITANTS OF THE NEW NEIGHBOURHOODS CAN ENJOY OPTIMAL INTERMODAL CONNECTIONS WHEN USING OUR MULTIMODAL PUBLIC TRANSPORT SERVICE."

5.5 CIVIL SOCIETY DRIVEN APPROACH

MOBILITY
Collaborative development
and services

CITY REGION
Globally networked in
transnational regions

NEIGHBOURHOOD
Community organization
in the neighbourhood

The civil society-driven scenario is characterized by civil society initiatives started by individuals and groups who think of CAM in terms of local needs and develop corresponding services. They are pioneers of technology-, sustainability- and sufficiency-driven change in mobility. Together they trial strategies for traffic avoidance while putting the active mobility of people and goods front and centre. Connected and automated vehicles are almost exclusively used in the context of sharing and at low speeds: ride sharing generally dominates passenger mobility. Small, automated bots transport goods in the region and also call into question transport-intensive, industrial production and distribution logistics.

In light of multiple crises caused by an environmentally destructive economic system and a loss of trust in politics, this scenario envisions a radical social transformation. It is based on a conviction that a socio-ecological reorganization of society is necessary; it incorporates various dynamics within the transformation debate and new designs for the "good life". At the core of this transformation is the notion of sufficiency and the principle of communality.

Research funding as part of education policy, knowledge creation and knowledge transfer takes place in global networks. Ideas are primarily implemented at local level. The main focus is learning skills locally that will contribute to the development and use of digital technologies in the transportation system. Civil society is increasingly able to improve new technologies and communication services (civic tech, open data, open source). At the neighbourhood level, and mostly on the edge of town and beyond, modern-day localism and communitarianism movements take centre stage.

It is assumed that antagonistic and transformative social movements are reflected in the space, meaning that the heterogeneity of different social groups is expressed spatially. Productive spaces and experimental areas emerge in places with low and medium urban density. What is produced in the city region and the resources locally available are distributed or shared locally via automated transporters. Furthermore, it is assumed that a strong ideological connection to a place would emerge as a spatial characteristic, which on occasion might seem socially exclusive. In the neighbourhood and in the streetscape, there is a preference for spaces that admit a wide range of uses and remain flexible.

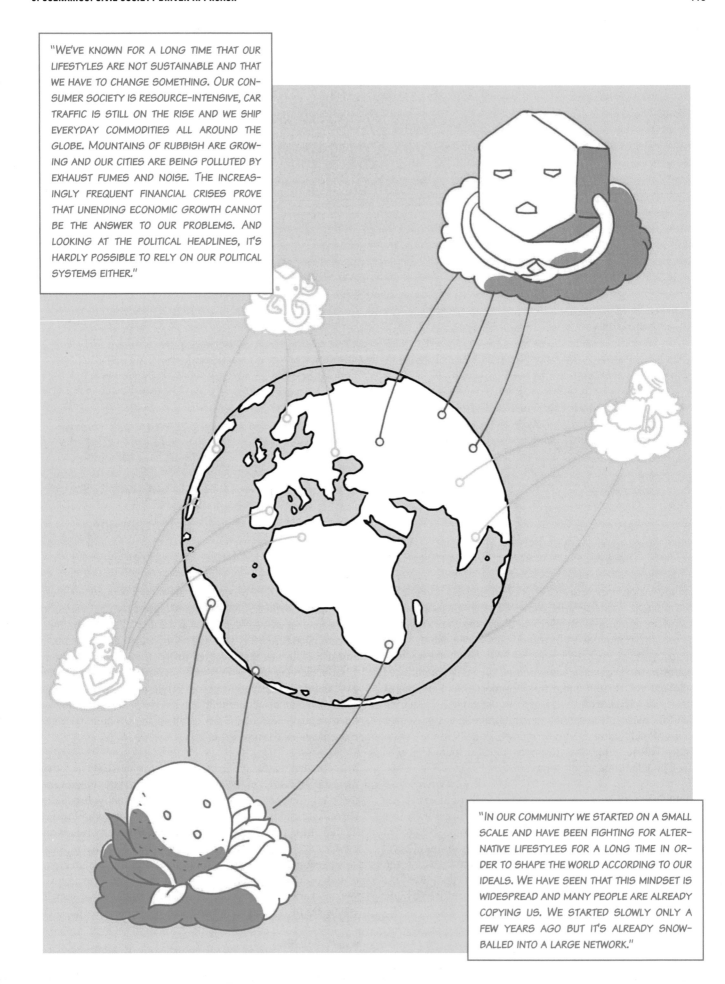

"WE'VE KNOWN FOR A LONG TIME THAT OUR LIFESTYLES ARE NOT SUSTAINABLE AND THAT WE HAVE TO CHANGE SOMETHING. OUR CONSUMER SOCIETY IS RESOURCE-INTENSIVE, CAR TRAFFIC IS STILL ON THE RISE AND WE SHIP EVERYDAY COMMODITIES ALL AROUND THE GLOBE. MOUNTAINS OF RUBBISH ARE GROWING AND OUR CITIES ARE BEING POLLUTED BY EXHAUST FUMES AND NOISE. THE INCREASINGLY FREQUENT FINANCIAL CRISES PROVE THAT UNENDING ECONOMIC GROWTH CANNOT BE THE ANSWER TO OUR PROBLEMS. AND LOOKING AT THE POLITICAL HEADLINES, IT'S HARDLY POSSIBLE TO RELY ON OUR POLITICAL SYSTEMS EITHER."

"IN OUR COMMUNITY WE STARTED ON A SMALL SCALE AND HAVE BEEN FIGHTING FOR ALTERNATIVE LIFESTYLES FOR A LONG TIME IN ORDER TO SHAPE THE WORLD ACCORDING TO OUR IDEALS. WE HAVE SEEN THAT THIS MINDSET IS WIDESPREAD AND MANY PEOPLE ARE ALREADY COPYING US. WE STARTED SLOWLY ONLY A FEW YEARS AGO BUT IT'S ALREADY SNOWBALLED INTO A LARGE NETWORK."

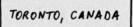

"WE FIND IT GREAT THAT PEOPLE ALL OVER THE WORLD ARE EXPERIMENTING. THE INTERNET AND NEW TECHNOLOGIES GIVE US THE OPPORTUNITY TO LEARN FROM INITIATIVES ON THE OTHER SIDE OF THE GLOBE AND TO COMMUNICATE WITH THE PEOPLE INVOLVED.

THAT COMMUNICATION IS IMPORTANT TO US BUT BEING SELF-ORGANIZED ALSO REQUIRES PROXIMITY, AND A SENSE OF COMMUNITY IN YOUR LOCAL AREA IS CRUCIAL FOR THAT. IT'S COMPLETELY CHANGED OUR NEIGHBOURHOODS.

FOR EXAMPLE, OUR COMMUNITY'S SUPPLY INFRASTRUCTURE WORKS WELL IN GREEN AREAS, TO OUR GARDENS OUTSIDE THE NEIGHBOURHOOD. ELSEWHERE, ROUTES ARE BEING EXTENDED. OUR INTERESTS ARE SO DIVERSE THAT YOU NO LONGER NEED ALL THE INFRASTRUCTURE, BUT ONLY THAT WHICH IS IMPORTANT FOR THE RESPECTIVE COMMUNITY."

"IT STARTED WITH US TRYING TO UNDERSTAND AND TACKLE OUR EVERYDAY PROBLEMS IN THE NEIGHBOURHOOD. WE FOUND SOME GREAT SOLUTIONS – THAT'S OUR DEFINITION OF INNOVATION. IN HOUSING, IN THE COMMUNITY AND IN MOBILITY, WE'VE DEVELOPED SERVICES THAT REFLECT OUR LIFESTYLES. FOR SOME ROUTES I HAVE TO TAKE A SHUTTLE, AND I LOVE RELAXING AND ENJOYING THE QUIET DURING THE JOURNEY. THAT'S WHY WE NETWORKED HERE IN THE NEIGHBOURHOOD: WE WANTED TO DEVELOP A SHUTTLE WITH A DESIGNATED QUIET AREA.

WE WERE ABLE TO ACHIEVE THAT TOGETHER WITH GLOBALLY NETWORKED SOFTWARE ENGINEERS AND OTHER SPECIALISTS IN CONNECTED AND AUTOMATED DRIVING, SO WE NOW OPERATE A VEHICLE THAT CORRESPONDS TO OUR INTERESTS AND IDEAS. WE WERE ABLE TO BUILD ON A FOUNDATION OF OPEN DATA, PEER-TO-PEER SERVICES AND OTHER DO-IT-YOURSELF INITIATIVES THAT HAD ALREADY IMPLEMENTED SIMILAR PROJECTS. IT HAS MEANT THAT WE CAN ORGANIZE LARGELY INDEPENDENTLY OF THE GLOBAL GOODS AND FINANCIAL MARKET."

"PEOPLE REACTED VERY SCEPTICALLY AT FIRST. OBVIOUSLY, IT REQUIRES LONG-TERM DEDICATION, TRUST AND SOLIDARITY. WE NEED PEOPLE WHO PAY THEIR MEMBERSHIP FEES RELIABLY AND FEEL RESPONSIBLE FOR MAINTENANCE. DEALING WITH PEOPLE WHO DON'T SHARE OUR INTERESTS AND DON'T ABIDE BY OUR RULES IS VERY CHALLENGING. OUR SERVICE PROVISION IS AIMED AT LIKE-MINDED PEOPLE, NOT PEOPLE ON A DIFFERENT WAVELENGTH. WE BELIEVE THAT SOME PEOPLE ARE BETTER SERVED BY OTHER ASSOCIATION."

"WE HAVE TO SPREAD NEW VALUES THAT SUPPORT A RESPONSIBLE APPROACH TO OUR NATURAL RESOURCES.

THAT ALSO MEANS THAT WE CRITICIZE THE CURRENT LIFESTYLES OF THE MASSES, BECAUSE THEY JUST AREN'T SUSTAINABLE. THE MEDIA GIVES A FALSE PICTURE OF THE GOOD LIFE, BASED ON THE IDEAL OF LIVING IN THE FAST LANE, WHICH MEANS SHORT AND LONG-DISTANCE FLIGHTS ON BUSINESS JETS TO THE FINANCIAL CENTRES OF GLOBAL CITIES, TO PRIVATE ISLANDS OR TO EXPENSIVE HOLIDAY HOMES. BUT WHAT WE REALLY NEED IS SLOW LIVING AND AN EVERYDAY LIFE THAT IS ROOTED IN THE LOCAL AREA."

"THE FIRST THING WE DID WAS TO APPROPRIATE PARKING SPACES. AT FIRST DRIVERS WERE FURIOUS AND HAD LITTLE UNDERSTANDING FOR OUR ACTIVITIES. IT TOOK SOME TIME FOR THE DRIVERS' PROTESTS TO SUBSIDE. BUT MUNICIPAL POLITICIANS AND URBAN PLANNERS HAD TO RESPOND TO OUR DEMANDS AND CREATE ARCHITECTURAL STRUCTURES SO THAT WE COULD GO TO THE PHARMACY OR SCHOOL BY FOOT OR BY BIKE. WE SHOULD ONLY GET INTO A CAR AND PRODUCE MOTORIZED TRAFFIC IN EXCEPTIONAL CIRCUMSTANCES. THAT'S OUR IDEA OF SUSTAINABLE MOBILITY."

"POLITICIANS AND THE ADMINISTRATION HAVE LARGELY RECOGNIZED THAT WE CITIZENS ARE BEST SUITED TO SOLVING OUR OWN PROBLEMS. WE WOULDN'T HAVE BEEN ABLE TO DO IT WITHOUT THE PUBLIC SECTOR'S TRUST IN US AND THE NEW PARTICIPATIVE ATTITUDE. HOWEVER, WE STILL HAVE TO COME TOGETHER TO SEARCH FOR NEW, GOOD LIFESTYLES AND GET INVOLVED IN AN ONGOING PROCESS OF POLITICAL NEGOTIATION WITH DIVERGENT OPINIONS. FOR EXAMPLE, INFRASTRUCTURE PROVISION DOESN'T WORK EVERYWHERE. IN THIS REGARD, WE SOMETIMES STILL NEED SUPPORT FROM THE STATE. IT ACTUALLY WORKS REALLY WELL IN OUR COMMUNITY, BUT MOBILITY SERVICES BEYOND OUR AREA ARE NOT REALLY VERY IMPORTANT TO US."

"COMMUNITY THINKING HAS ACHIEVED IN-CREDIBLE THINGS IN OUR NEIGHBOURHOOD AND GIVEN RISE TO NEW FORMS OF SOLIDARI-TY AND SOCIAL AND ECONOMIC SECURITY. BUT JUST RECENTLY, OUR SHUTTLE WAS BROKEN AND I HAD AN IMPORTANT APPOINTMENT. BE-CAUSE I'M NOT A MEMBER OF ANOTHER MO-BILITY ASSOCIATION, I CAN'T SPONTANEOUSLY ARRANGE ANOTHER RIDE. WHAT'S MORE, I DON'T KNOW THE ACCESS REQUIREMENTS AND IN THE WORST CASE I'D HAVE TO SHARE A RIDE WITH A BUNCH OF TEENAGERS. SINCE I'M USED TO TRAVELLING IN MY QUIET SHUTTLE, PEOPLE TALKING LOUDLY DURING THE JOURNEY RE-ALLY AGGRAVATE ME. WHAT ABOUT THOSE WHO DON'T FIND ANY LIKE-MINDED PEOPLE TO START AN ASSOCIATION WITH? I THINK THEY HAVE A HARD TIME IN OUR COMMUNITY. BUT THEY CAN ALWAYS GET INVOLVED AND OFFER THEIR OWN SERVICES."

5.6

TABULAR COMPARISON OF THE THREE SCENARIOS

	MARKET-DRIVEN	POLICY-DRIVEN	CIVIL SOCIETY-DRIVEN
APPROACH			
Characterization	free market	steering state	strong civil society
Main actor	private companies	state	civil society
DRIVERS AND OBJECTIVES			
Main aim	efficiency, profit	public interest, environmental and climate protection, health	sufficiency, self-determination, sustainability
CORE			
Transport policy aim	efficiency	modal shift	traffic prevention
Urban policy model	competitive city	socially inclusive city region	participative city region
Underlying conditions	deregulation of the mobility market and liberal economic conditions	systematic approach to shaping the mobility market and services	opening up the mobility market to civil-society initiatives
Role of the state in the mobility market	weak: provides basic services	strong: shapes the mobility market	passive: makes initiatives possible
Financing models	use or valorization of data	subsidization using public funds	mobility as a public good (commons-based)
MOBILITY MARKET			
Structure	oligopoly of private actors	state-steered oligopoly	multisectoral networks
Competition and alliances	strategic alliances between technology producers and mobility service providers (aviation)	private actors involved under strict terms (licences)	alternate cooperation and redundancy

	MARKET-DRIVEN	POLICY-DRIVEN	CIVIL SOCIETY-DRIVEN
NEW MOBILITY: MOBILITY AS A SERVICE, SHARING, PUBLIC TRANSPORT (CHAP. 3.3)			
MaaS integrator	private	public	public-private-people partnerships
Obligation to carry	no	yes	no
Service vs. price differentiation	+++	+	++
Operation financing	users: +++ tax money: +	users: + tax money: +++	users: ++ tax money: ++
Horizontal integration	CAT as a door-to-door service; micromobility and conventional car sharing in CAM-incompatible areas where possible	conventional public transport as the backbone; CAVs expand the public transport services offered; planned availability of modes of transport	CAV is only a door-to-door solution where active mobility reaches its limits; integrates a wide choice of different providers
Vertical integration	integration of all services, "premium accounts" (Level 4)	integration of policy aims (Level 4)	integration varies (Levels 2–3)

RELEVANCE OF MARKET PLAYERS

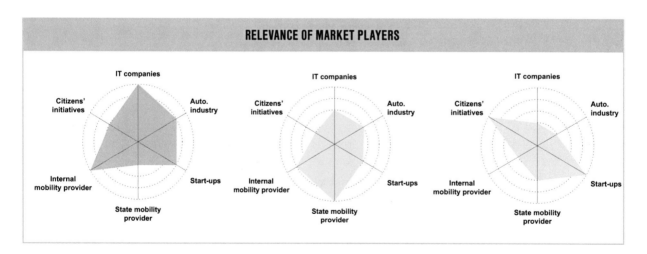

TRANSPORT AVAILABLE

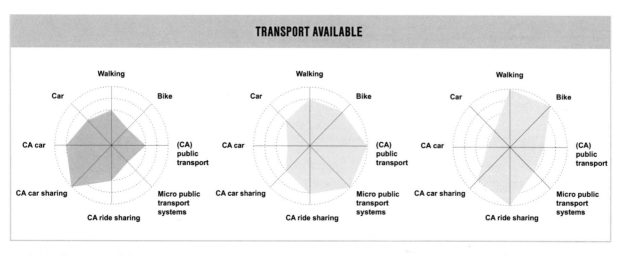

GOVERNANCE

	MARKET-DRIVEN	POLICY-DRIVEN	CIVIL SOCIETY-DRIVEN
ECONOMY			
Pricing, fiscal policy	• Mobility pricing	• Moderate mobility pricing • Taxation on VA cars • Licences or provision contracts	• Mobility pricing
ENFORCEMENT			
Legal measures, regulatory policy	• Focus on security standards • Compulsory insurance	• Bans on connected and automated cars, limitation of empty journeys • Restrictive land policy	• Driving bans dependent on "streetscape compatibility"
EDUCATION			
Communication, information	• Advertising	• Nudging (influencing behaviour)	• Skill development
ENGINEERING			
Planning, infrastructure construction and operation	• Adaptation of road infrastructure to CA compatibility	• Expansion of the public transport system and of mobility hubs	• Speed reduction for CAVs and construction of micro hubs

URBAN SPACE

	MARKET-DRIVEN	POLICY-DRIVEN	CIVIL SOCIETY-DRIVEN
TRANSPORT HUBS			
Hierarchical concept	flat	multilevel	flat
Functional integration	consumption	social and education institutions	comprehensive
Access	semi-public space	public space	public spaces
Local integration	+	+++	++
Design quality	+++	++	++

	MARKET-DRIVEN	POLICY-DRIVEN	CIVIL SOCIETY-DRIVEN
PUBLIC SPACE			
Streetscape quality	polarized, commercialized	hierarchization, controlled	appropriated, different uses
Separation principle	++	+++	+
Mixing principle	++	+	+++
Ground-floor use	spatially concentrated	curated	diverse

	MARKET-DRIVEN	POLICY-DRIVEN	CIVIL SOCIETY-DRIVEN
CITY REGION			
Spatial structure (density, centrality, accessibility)	strong centre, suburbanization and decentralization in surrounding areas	mixed-function settlement centres and polycentric structure in urban agglomerations	regional, intertwined islands (mosaic)
Function of the big city and small town centrality	city as economic and steering centre	regional integration, polycentricity	local centres, formation of enclaves
Land management	business-led land management	land management clearly specified by the public sector	land management regulated via the community

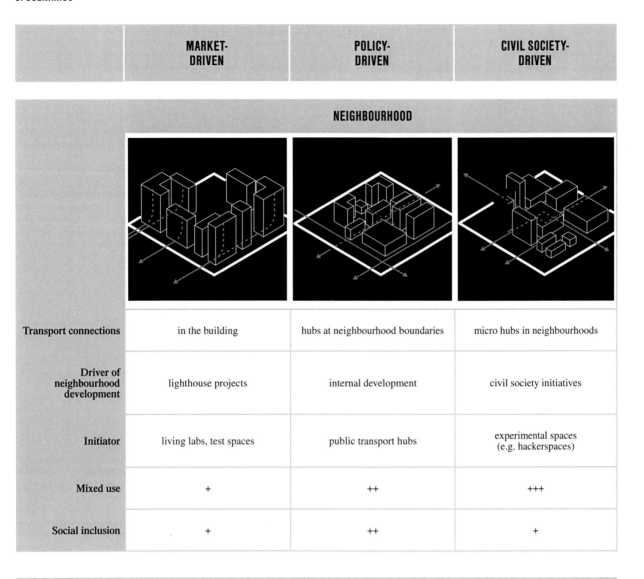

	MARKET-DRIVEN	POLICY-DRIVEN	CIVIL SOCIETY-DRIVEN
NEIGHBOURHOOD			
Transport connections	in the building	hubs at neighbourhood boundaries	micro hubs in neighbourhoods
Driver of neighbourhood development	lighthouse projects	internal development	civil society initiatives
Initiator	living labs, test spaces	public transport hubs	experimental spaces (e.g. hackerspaces)
Mixed use	+	++	+++
Social inclusion	+	++	+

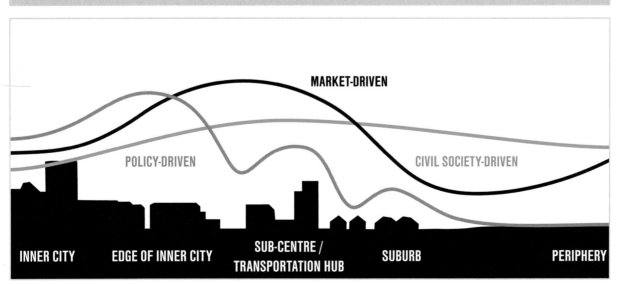

LAND USE DYNAMICS

MARKET-DRIVEN

POLICY-DRIVEN   CIVIL SOCIETY-DRIVEN

INNER CITY   EDGE OF INNER CITY   SUB-CENTRE / TRANSPORTATION HUB   SUBURB   PERIPHERY

Source: AVENUE21

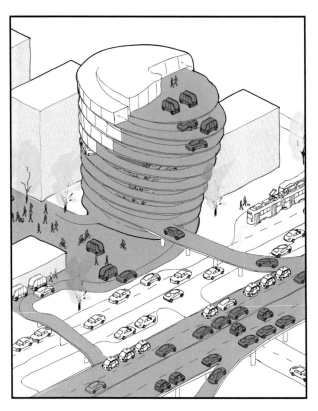

← INTERNET OF THINGS NEIGHBOURHOOD

The mobility hub in the market-driven scenario functions as a gateway into and out of the neighbourhood. Located alongside a main transport route, the mobility and property markets are completely integrated into the location development. Companies and residents in the neighbourhood can access various mobility services and vehicle types as the situation requires. Public space is privatized and developed as shared space.

CONNECTED AND AUTOMATED MICRO HUBS →

In the civil society-driven scenario, various micro hubs are located in the ground floors of buildings. The boundaries between streetscape and building interiors are fluid. The neighbourhood can only be accessed at low speeds, which change according to the current use of the space. At the edge of the neighbourhood, large-capacity public transport (partly) takes over connections to the city region.

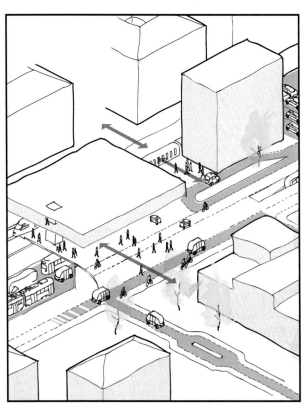

← CAT-ASSISTED TRANSIT-ORIENTED DEVELOPMENT

In the transit-oriented development of the policy-driven scenario, the multimodal hub is in the centre of the neighbourhood. In line with the seamless transport approach, wherever possible efforts have been made to remove barriers between everyday life in the neighbourhood and public transport access. Pedestrians and cyclists can directly access the tracks and stops/stations, while private CAVs can be parked a short distance away.

Illustration: Alexander Diem

5.7

STAKEHOLDERS' ASSESSMENT OF THE SCENARIOS

In order to find out how the scenarios developed in the project are evaluated by various stakeholders, focus groups were set up with experts from different areas (city-regional governance, planning, mobility service providers, technology companies and scientists). From these, conclusions were inferred for the subsequent scenario development process. Furthermore, the discussions in the focus groups were used to gain experience of how the scenarios could be implemented and to discuss the impact of CAT with a broader expert audience as well as with the lay public.

As the use and application of CAT lie in the future, there is by and large a lack of practical knowledge, and as CAM as a research topic is highly complex, the methodological approach faces particular challenges. To address this, classic methods of empirical social research and futurology were combined. There are interesting starting points in this regard in the field of critical utopian action research, citizen science approaches, speculative and visual sociology and design thinking approaches (Levitas 2010, Robinson et al. 2011, Husted/Tofteng 2014, BuroHappold Engineering 2016, Freudendal-Pedersen et al. 2017). The aim was to find out various stakeholders' everyday experiences and criticism of the current mobility system as well as their imagined future utopias related to CAV in order to investigate the potential impacts of CAM (Levitas 2010: 542).

In total, three focus groups were organized; in each of them one scenario and related visions of the future were discussed. After presenting the scenarios, the potential impacts, opportunities and risks were debated. This was intended to stimulate the participants' imaginations regarding the possibilities of CAM; building on that, the various application scenarios were then discussed. Finally, the group of experts identified what action was deemed necessary. The identified opportunities, risks and need for action are summarized in Figs. 5.7.1 to 5.7.3.

The results show that due to their different interests and viewpoints as well as the complexity of the object of research, there is a very diverse body of knowledge and divergent opinions on CAT. As communication tools, the scenarios can prompt a discussion about possible futures and associated needs for action. In the context of collaborative planning processes, this can enable an early definition of various solutions in the area of mobility as well as settlement and neighbourhood development in connection with CAM. Collaborative processes also contribute to individual and collective learning, which can make decision-making by local stakeholders more flexible (Innes/ Booher 2010). Moreover, there is no universal solution to many challenges posed by CAT. Discussing possible scenarios with different social groups can make the negotiation process regarding future developments more transparent and support the decision-making process.

Figure 5.7.1: **Opportunities identified by the stakeholders in the three scenarios**

MARKET-DRIVEN	POLICY-DRIVEN	CIVIL SOCIETY-DRIVEN
• The city profits more than the urban fringe. • The competition between international actors on local mobility markets encourages diverse and attractive mobility services. • When pricing pressure between various CAM providers increases, the services become more cost-effective. • Interesting experiments can be expected by and as a result of CAM.	• Potentials to mitigate the urban-rural divide lie predominantly in rural areas. • The inclusion of certain social groups is furthered by CAM.	• Better accessibility of sparsely populated areas with CAVs reduces the urban-rural divide. • Incorporating citizens' wide-ranging knowledge in innovative processes fosters personalized mobility services. • Sharing in the field of mobility could lead to a new, growth-critical economic system. • Community-organized mobility strengthens identification with the neighbourhood.

Figure 5.7.2: **Risks identified by stakeholders in the three scenarios**

MARKET-DRIVEN	POLICY-DRIVEN	CIVIL SOCIETY-DRIVEN
• If pricing pressure between different CAM-providers increases, market adjustments will happen in the medium term. • The commercialization of travel time will become a driver of further traffic growth. • Without steering, considerable cost pressure on infrastructure and development pressure on settlement structure (urban sprawl) are to be expected. • If the spatial availability of CAT reacts to demand, socio-spatial inequalities could be exacerbated. • A monopoly or oligopoly attitude cannot be ruled out, which would result in prices being raised. • There will be "new winners" and "new losers" on the mobility market.	• CAM could lead to a rigid organization and hierarchy at a political level. • CAM could also be instrumentalized in various ways for party-political purposes. • Public sector resources are limited and budgets might not suffice to achieve policy and planning aims. • Major sanctions during the introduction of CAVs are difficult to enforce at a policy and planning level.	• It is not possible to guarantee citizens' long-term involvement. • Isolated solutions at different stages of development will emerge, which will lead to inequalities unless the market or state intervenes. • How will highly personalized mobility services be standardized? • The transferability of civil society-driven projects might prove difficult. • Civil society-driven CAM projects might develop a close proximity to market interests in the long term. • The pressure on settlement structure (urban sprawl) might increase with weak overarching and strategic planning.

Figure 5.7.3: **Needs for action identified by the stakeholders in the three scenarios**

MARKET-DRIVEN	POLICY-DRIVEN	CIVIL SOCIETY-DRIVEN
• State regulations should offset unsustainable market effects. • Cities competing for CAV test areas might lead to security standards being lowered; this should be prevented by appropriate state regulations. • Route licences could be introduced to ensure operation of the route even with CAVs. • The private public transport providers must enter into service agreements with the public sector. • Many stakeholders would have to be involved in designing a regulatory system that controls CAT traffic steering and licensing for public transport and delivery services.	• Specific services should be supported: micro public transport systems, shared on-call taxis, car sharing. • Mobility services should be organized via a platform coordinated by the city. • CAV-compliant "mobility points" must be specially planned to create transfer zones for passengers to change to micro mobility and active mobility. • In order to avoid further urban sprawl as a result of CAM, strong land policy tools are necessary. • In less easily accessible areas, the public sector must guarantee adequate mobility services using CAVs. • In order to appropriately steer the introduction of CAVs, action should be taken early while also considering the long-term perspective. • The city's possibilities for action consist primarily in preparing and operating infrastructure. • The regulatory system for traffic steering or licensing must be drawn up by policymakers and planners. • The information on the impact of CAT must be shared with the communities in as comprehensive a way as possible.	• In less easily accessible areas, the public sector must guarantee adequate mobility services using CAVs. • Appropriate participation in the implementation of CAM projects is thus far lacking and should be enabled and encouraged by the city. • Education initiatives and incentive schemes should be used to transform current-day traffic behaviour. • Experimental spaces should be created for bottom-up initiatives.

5.8

IN-DEPTH CONSIDERATION OF SPATIAL DYNAMICS IN THE LONG LEVEL 4

In this section, the three scenarios will be discussed in more depth and evaluated regarding the impacts of connected and automated transport in SAE Level 4 in terms of accessibility, site selection and reclaiming parking spaces at various spatial scales. This is intended to provide support for urban and mobility planning as it prepares for the challenges that will arise during the Long Level 4 with regard to steering and planning in the city region. Such a practice-oriented approach has thus far been lacking.

The strength of the normative narrative scenarios presented here consists in their portrayal of various conceivable developments and circumstances, which makes them both transparent and concrete, in order to ultimately illustrate the need for action on the path to a desirable future. According to the preambles of the strategies for development in European city regions, the aim is to create liveable and compact cities with a functional mix that are characterized by high-quality public space and that integrate CAT in an agreeable way. As this integration takes effect at various levels, the following analysis will differentiate between impacts on the public space, the neighbourhood and the city region.

5.8.1. PUBLIC SPACE

The streetscapes in the market-driven scenario are designed on the one hand for efficient traffic flow and on the other for optimum economic use. This leads to a polarization of public space: high-quality spaces increase the value of neighbouring properties and encourage gentrification (see Fig. 5.8.1, bottom left), whereas along efficiently designed arterial roads residents suffer from high levels of noise and air pollution and a greatly reduced quality of public space (see Fig. 5.8.1, bottom right).

The financing necessary to convert existing and build new digital infrastructure is obtained via public-private partnerships (PPPs). The owners of surrounding buildings participate in the form of modified business improvement districts (BIDs) in order to guarantee safety, cleanliness and control and to establish shared spaces in residential and office areas, where CAVs take care of the last mile at low speeds. Some pick-up and drop-off points for automated car sharing are located inside these buildings – in repurposed underground car parks, in formerly vacant shops or as newly created mobility

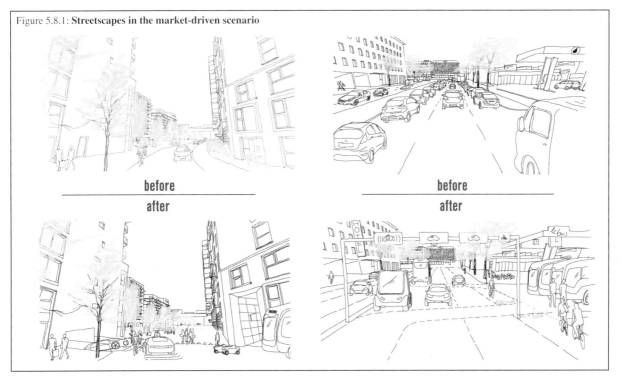

Figure 5.8.1: **Streetscapes in the market-driven scenario**

before
after

before
after

Source: AVENUE21

points where it is possible to choose between different kinds of automated vehicle and new urban mobility services (such as e-scooters).

The bulk of the traffic is concentrated along the routes that are designed for CAVs to be used at relatively high speeds. Here, the vehicles are directed into different lanes to simplify driving for the CAVs. These routes form their own network within the existing road infrastructure and range from motorways to city streets. The separation effect along these transport routes is extremely high as a result of the small distances between automated vehicles and the limited possibilities to cross such streets.

In the policy-driven scenario, the aim is to expand the public transport network by adding CAT and to improve the quality of public space. That also means that the streets or zones where the use of CAVs would negatively impact the quality of the streetscape are declared CAV-free areas. In order to reclaim public space, the city makes public transport more attractive by developing mobility platforms and apps. Due to reduced car traffic, former parking spaces can be repurposed to the benefit of cycle paths, public seating or playgrounds. As in the market-driven scenario, connected and automated vehicles or their use in streets in dense urban locations is only possible with drawbacks: distance and barriers between CAVs and other road users are necessary, which limits the use of space (see Fig. 5.8.2).
In the civil society-driven approach, public spaces

are transformed by their use. Speeds are dramatically reduced. The reclaimed land, which – particularly in previously car-friendly areas – is considerable, is now used for functions for which land is currently needed. Automated mobility services connect and interlink productive parts of public space and use ground-floor areas. However, the heterogeneity of the streetscape, the functional mix and variety of vehicles also have negative impacts on road safety. In most cases, the reduced speeds have contributed to a radical transformation of the streetscape and supported urban transport planners' current strategies to introduce more 20 and 30 km/h zones or to make inner cities car-free (see Fig. 5.8.3).

PRELIMINARY CONCLUSIONS AND COMMONALITIES

Fundamental contradiction: high-quality public space and CAM at higher speeds
Every scenario reveals a contradiction between vibrant public spaces (with different transport users and ground-floor uses) and CAVs at higher speeds. Although higher speeds and larger deployment areas can be realized in cities with the aid of infrastructure measures, this is always at the cost of existing streetscape quality.

Public space is only reclaimed when accompanied by traffic prevention measures
When writing the scenarios, it became clear that during the transition period, parking spaces in the streetscape do not become obsolete to the extent anticipated purely as a

Figure 5.8.2: **Inner-city streets in the policy-driven scenario**

before

after

Source: AVENUE21

Figure 5.8.3: **Street on the edge of town in the civil society-driven scenario**

before

after

Source: AVENUE21

result of CAV introduction. Instead, the main measures that are required are behavioural changes, a wide range of sharing services, higher-quality public transport and infrastructure and spatial structures that support active mobility. In every scenario, additional planning and/or steering measures are necessary to combine the aims of the new transportation revolution with the use of CAVs.

3.8.2. NEIGHBOURHOOD

The pioneering neighbourhoods in the market-driven scenario have a global focus. They are in keeping with the expectations of a highly mobile clientele; locations are compared and selected internationally. In IoT neighbourhoods, automated mobility services embedded in digital ecosystems become a fundamental part of digitalized everyday life. This leads to specific places on the edges of the city near motorways being preferred, because here a kind of neighbourhood development is possible in which CAVs can travel without limitations. Starting from these locations, the transnational transport network used for passenger and freight transport is easily accessible. In these IoT neighbourhoods, businesses, residents and workers all benefit, while less easily accessible neighbourhoods and parts of the city undergo economic stagnation or regression.

In the policy-driven scenario, policy and planning advance neighbourhood development with targeted infrastructure projects. This begins with local mobility accessibility planning and includes not only various vehicle types and transport networks, but also targeted location development in close proximity to transport hubs (schools, nursery schools, administrative offices, retail areas, etc.). The aim is to create dense neighbourhoods with a functional mix. Especially in inner-city areas, CAT is treated quite restrictively and is developed to supplement public transport in the context of expanded sharing services. High-quality public transport is further concentrated and follows a specific route network. Connected and automated shuttles and ride-sharing services add to the transport network and are not restricted to any established route network in order to close gaps in the existing public transport provision.

Neighbourhood development in the civil society-driven scenario relies on the initiative of the local residents. To this end, discussion spaces are created in which residents' existing knowledge of everyday problems and needs is utilized and feeds into designs for mobility and spatial development. The boundaries between inside and outside, green areas and the streetscape are fluid. As a result of the participative approach to neighbourhood development, people's identification with the area improves. However, that does not exclude the possibility that in certain constellations very specific interests regarding the reorientation of transport or the design of public space in the neighbourhood prevail.

PRELIMINARY CONCLUSIONS AND COMMONALITIES

In the neighbourhoods, Level 4 pockets develop, i.e. urban subspaces with a network of roads/infrastructure suitable for the use of CAVs (see Fig. 5.8.4). They emerge in all scenarios and are consciously planned, technology-dependent or initiated by local interests. This phenomenon has been identified elsewhere (European Commission 2017: 96).

In the market-driven scenario, such "exclaves" are the result of unlimited technological possibilities and targeted, highly suitable, planned neighbourhoods. This leads in particular to sites near motorways becoming drivers and transport hubs in a privatized, transnational network. With the policy-driven scenario, it becomes clear that it can be expedient for planning aims to restrict access or ban CAVs in zones where the quality of public space would suffer from the arrival of CAVs. Whether steered by policy and planning or dependent on technology, assuming a Long Level 4 means that fractures in the transport system and urban fabric can be expected, which will lead to new challenges at neighbourhood level. It follows that:

High CAT accessibility will be an important criterion for companies when choosing a location
It is conceivable that e.g. offline retail, but also offices and production sites will benefit from locations where automated delivery and operation is possible, whereas places without that infrastructure will regress. That would mean additional disadvantages and increased pressure to act for areas close to the inner city like shopping streets and neighbourhood centres.

Road safety risk
At the boundaries of these islands, motorized private transport has to be transferred from the automated technical system to human drivers, which might present a considerable safety risk (Hellåker et al. 2019). These interfaces must be under technological surveillance and an orderly transfer must be ensured (e.g. by means of geofencing or by the automated driving system; Stark et al. 2019). Furthermore, other road users would have to be informed about the mode currently being used by the vehicles and adapt their behaviour accordingly.

Neighbourhood development around mobility hubs
The presented three scenarios addressed different conceivable typologies of mobility hubs (see Figs. 5.8.5, 5.8.6) and their integration into the neighbourhood. In the market- and policy-driven scenarios, possibilities for targeted location development around the new mobility hubs are identified. In the market-driven scenario, as discussed above, new mobility hubs appear particularly on the edge of the city and function as an interface between the motorway network and the city region (see the typology on the left of Fig. 5.8.5). In the policy-driven scenario, the concept of transit-oriented

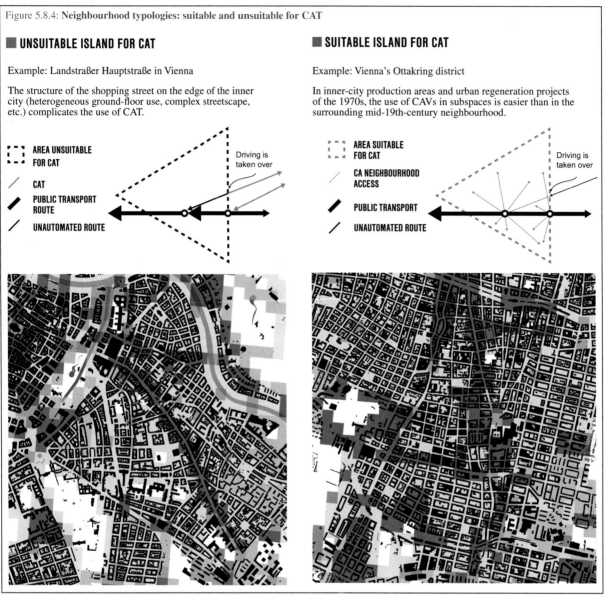

Figure 5.8.4: **Neighbourhood typologies: suitable and unsuitable for CAT**

Source: AVENUE21

development is taken further with CAT technology (see Fig. 5.8.5, centre). The civil society-driven scenario addresses the idea of local integrated micro hubs (see Fig. 5.8.5, right).

The typologies of possible CAT mobility hubs presented in the three scenarios were developed by means of forward projection based on existing neighbourhood development around transport hubs (see Fig. 5.8.6).

The neighbourhood around the transport hub in the market-driven scenario functions like a main railway station in an otherwise monomodal CAV-based transnational transport system. The policy- and civil society-driven scenarios feature different possibilities (and scales) for integrating a multimodal transport system at neighbourhood level.

5.8.3 CITY REGION

Under the laissez-faire policy and planning depicted in the market-driven scenario, urban sprawl happens rapidly and spreads along the edge of motorways. Here the advantages of CAT can be used on the long commutes (short to medium term: safety, efficiency, alternate use of travel time, convenience; long term: choice of company and residential locations). In location development, the mobility and property markets are intentionally combined. Different uses (logistics, manufacturing, housing, offices and retail) form clusters in areas that best correspond to their transport needs. Residential, business or industrial parks benefit from the lower price of land in the urban fringe. Areas with poor CA accessibility (predominantly the city centres) are deprived of functions and largely lose their significance as centres. In part, routes for automated vehicles emerge in inner-city areas, which primarily provide links with e.g. shopping streets (see Figs. 5.8.5, 5.8.7).

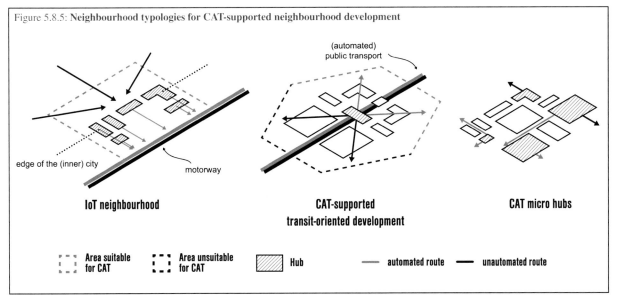

Figure 5.8.5: **Neighbourhood typologies for CAT-supported neighbourhood development**

IoT neighbourhood

CAT-supported
transit-oriented development

CAT micro hubs

(automated) public transport

edge of the (inner) city

motorway

| | Area suitable for CAT | | Area unsuitable for CAT | Hub | automated route | unautomated route |

A restrictive land policy prevents urban sprawl in the policy-driven scenario. In order to counteract increasing property prices in a growing city region, neighbourhoods are intentionally developed around mobility hubs, which are surrounded by strict settlement limits. In order to stop location quality moving to areas along motorways, a modal shift to (partly automated) public transport is encouraged and a city-regional logistics network is developed to strengthen offline retail. Alongside the rail network, which maintains its important role in the transport system, use is made of possibilities for automated public transport on the motorway (via bus rapid transit (BRT)). The most important task undertaken by CAT is providing access to low- and medium-density built-up areas in between major transport routes, which allows them to be better integrated into the public transport network (see Fig. 5.8.7).

The civil society-driven scenario presents a fundamental structural change starting primarily from the edge of the inner city. A tight network of different sharing services extends across the city region. For short-distance travel, public transport has become less important than active mobility, whereas public transport remains critical for accessing the region. Poorly planned areas of low and medium density are repurposed and unused spaces are made productive. Settlement and administrative boundaries are almost entirely dissolved. In order to avoid travel wherever possible, wares are produced in the region, the use of existing resources is increased thanks to

Figure 5.8.6: **Neighbourhood development around mobility hubs and historical models**

	TIME PERIOD	EXAMPLE	SPATIAL FOCUS	MAIN ACTORS	DRIVER
Capitalization of land	1980s	Broadgate (London)	railway station site, neighbourhood	rail companies	privatization of rail, neoliberal markets
Urban megaproject	1990s	Euralille (Lille)	(new) main station, adjoining neighbourhood	local and national governments	construction of high-speed trains, transnational structural policy
Transit-oriented development	2000–2010	Stedenbaan (Rotterdam–Den Haag)	several neighbourhoods along the regional rail network	coalitions of regional governments and public transport operators	expansion of regional rail networks, sustainable settlement development
IoT neighbourhood	Long Level 4	Fig. 5.8.5, left	neighbourhood near motorway exit	mobility service providers, IT companies	technological limitations of CAVs
CAT-supported transit-oriented development		Fig. 5.8.5, centre	neighbourhood along a public transport route	regional coalitions, PPPs	sustainable settlement development, freeing up centres
CAT micro hubs		Fig. 5.8.5, right	partial neighbourhood	bottom-up-initiatives, start-ups	differentiation of mobility behaviour

sharing services and automated shuttles combine passenger and goods transport. Buying local and a strong sense of community support decentralized structures of various density. Micro hubs become small-scale focal points of urban transformation. The function of centres is strongly defined by at times very specific group interests (see Fig. 5.8.7).

PRELIMINARY CONCLUSIONS AND COMMONALITIES

Redesign of regional and supra-regional transport
Regional and supra-regional transport changes dramatically as a result of the interplay of traditional and technologically upgraded vehicles:

- On motorways, it is of primary importance to coordinate moderately and highly automated transport (SAE Levels 2, 3 and 4).

- Current regional and supra-regional public transport services (bus, rail) risk losing their USP (conveying people), have to counter increasing cost pressure and face the challenge of existing in the face of convenient, personalized mobility services (see Fig. 5.8.8).

- When developing automated shuttle transport in areas between transport routes, it is important to ensure that urban sprawl in these areas is not encouraged.

- The global focus in the market- and civil society-driven scenarios shows that synergies (pooling several transport services) can be increased on mobility platforms by means of economies of scale (applying the service to other city regions or joint development). This is made clear by the example of the product range offered by Uber and open-source projects like comma.ai.

- More regionally focused mobility service providers or mobility platforms like those described in the policy-driven scenario are dependent on the economies of density continuing to exist as a result of high demand in the region.

Continued urban sprawl as the first spatial effect
Without countermeasures actively being taken by policymakers and planners, increasing urban sprawl as a result of CAT is probable at a very early point in the transition. A high degree of cooperation between the municipalities in a city region is imperative to mitigate the imminent rise in land use (Getzner/Kadi 2019). Reacting to this challenge will only be possible if the planning framework and taxation and incentive policies are fundamentally changed, new coalitions are established in settlement and transport planning and hence integrated regional planning is made possible (ÖROK 2015).

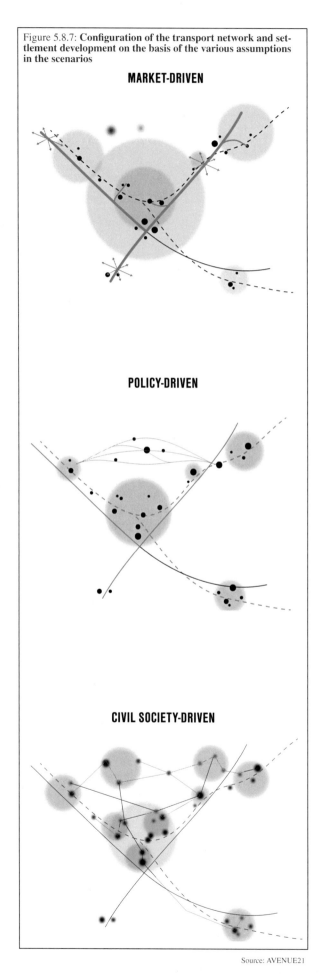

Figure 5.8.7: **Configuration of the transport network and settlement development on the basis of the various assumptions in the scenarios**

MARKET-DRIVEN

POLICY-DRIVEN

CIVIL SOCIETY-DRIVEN

Source: AVENUE21

Encouraging CAT in car-friendly areas
CAT's greatest potential can be exploited in areas with low and medium structural density, which would simultaneously make a significant contribution to the transportation revolution. By controlling the integration of CAT into the existing public transport service, it becomes possible to supply such areas with mobility services. New mobility hubs require space, more of which is available there than in the built-up inner cities (see Chap. 4.2). Both the policy- and the civil society-driven scenarios hint at how a change in transportation can be supported even in these settlement structures: by controlling their integration into and the expansion of the existing public transport network by means of CAVs and by densifying the area, increasing the functional mix and breaking up large units to create a "region of short distances" that is supplied and accessed by CAT at low speeds.

Level 4 Gap
The fundamental contradiction between high-quality public space and high-speed automated transport has already been mentioned. In Figure 5.8.8, it becomes clear that most European cities already have a relatively well-developed public transport network. The opportunity thus presents itself for European cities to take a separate path where multimodality might play an increasingly important role.

Planning task: various principles of CAT provision
When writing the scenarios, different CAT provision typologies were attributed to the various scenarios, which made the possibilities for shaping city-regional mobility planning clear. In the market-driven scenario, local and (trans-)national transport routes were directly connected via motorways (see Fig. 5.8.9.a). It is assumed that this form of transport provision favours individual forms of

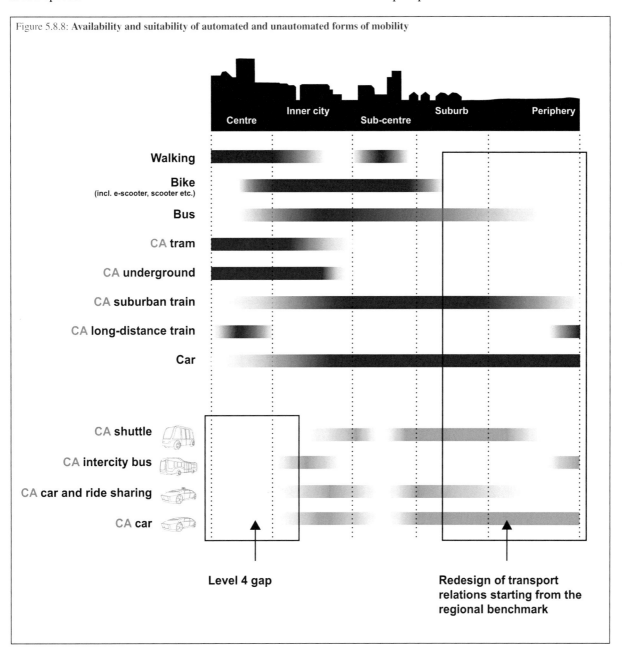

Figure 5.8.8: **Availability and suitability of automated and unautomated forms of mobility**

Figure 5.8.9: **Principles for CAT provision**

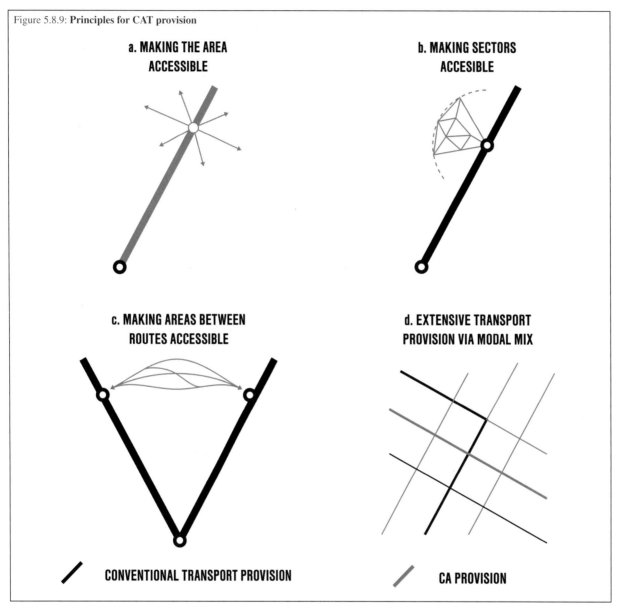

a. MAKING THE AREA
ACCESSIBLE

b. MAKING SECTORS
ACCESIBLE

c. MAKING AREAS BETWEEN
ROUTES ACCESSIBLE

d. EXTENSIVE TRANSPORT
PROVISION VIA MODAL MIX

CONVENTIONAL TRANSPORT PROVISION

CA PROVISION

Source: AVENUE21

mobility. The policy-driven scenario allows a strategically planned, multimodal CA public transport system to be developed that intentionally provides access to sectors and corridors and integrates them into a hierarchical system (see Figs. 5.8.9.b and c). The civil society-driven scenario addresses the mixture of automated and unautomated transport networks (see Fig. 5.8.9.d).

ACTION PLANS:

HOW CONNECTED AND AUTOMATED VEHICLES CAN SHAPE THE MOBILITY SHIFT

6

6.1

RE-EVALUATING THE POSSIBLE IMPACTS OF CONNECTED AND AUTOMATED VEHICLES IN THE CONTEXT OF A LONG LEVEL 4

Media outlets, as well as specialist engineering and science journals, have repeatedly been stating that connected and automated driving will soon be possible on all public roads. But in recent years, doubt has increasingly been cast on this supposed certainty. The analysis conducted as part of this research also suggests that the development of technologies for CA vehicles and infrastructures is moving more slowly than initially anticipated.

Although there has yet to be any empirical data on the speed of technological development, citizens' future levels of acceptance, progress in market penetration and the challenging implementation of the technology on roads with multiple types of user, it is vital that the policy planning stage be set now in order to help bring about the necessary shift in transport and mobility as well as promote the growth of sustainable transport and urban (district) development. In the debate on future mobility, CAT is considered to offer many benefits. This is based upon six key assumptions: CAT (1) reduces the number of road traffic accidents, (2) allows more efficient and effective traffic management, (3) enables fewer resources to be used and lower emissions, (4) reclaims public space, (5) provides inclusive individual transport to those with reduced mobility, and (6) allows more effective use of time while travelling (European Commission 2018, 2019). However, these potential outcomes will only occur under certain conditions and, in most cases, only in the medium to long term. Figure 6.1.1 outlines each of these theories in turn and compares them with arguments put forward by policymakers.

The assertion that CAVs will result in transport being avoided altogether has been refuted by most scenario-based studies. This shift might also have negative impacts for sustainable transport and urban (district) development plans due to the growing appeal of suburban locations. The argument that those with reduced mobility will once again be able to travel independently is only relevant to those groups who can walk as well as enter and exit a vehicle unaided. Whether and to what extent technologies will help strengthen social cohesion remains, for now at least, a mere normative hope that, at best, concerns marginal groups.

Although such scepticism has had little impact on the ongoing development of connected and fully automated vehicles and the required infrastructure, an increasingly critical stance towards the narrative that has accompanied and promoted this trend is, to some degree, necessary (see Chap. 4.6). On the other hand, it is vital that relevant strategies be developed in good time – both at the local and regional level – to ensure the best possible integration of CAT into existing transport, mobility and urban (district) and regional development goals.

Figure 6.1.1: **Proposed benefits of connected and automated transport and their practical and temporal constraints**

ARGUMENTS FOR FULLY AUTOMATED VEHICLES (LEVEL 5)	LIKELY IMPACTS IN A LONG LEVEL 4
In future, it will be possible to largely avoid **road traffic accidents** and to considerably reduce their negative impacts.	With CAVs showing a rising rate of penetration, together with a level of connectivity that allows real-time vehicle control, these objectives can be reached. In the interim, however, there will be a period in which vehicles on our roads have vastly different levels of automation and possible areas of application. This will cause passengers and other road users to feel anxious about how effective the vehicle will be and whether it is under automated control, thus increasing the risk of accidents, at least during the initial phase.
Traffic flows can be managed more efficiently and effectively to largely avoid congestion and to ensure better use of transport infrastructure capacity. This means less investment in infrastructure expansion will be required even if traffic volumes continue to rise.	Simulation studies generally assume that every section of road is suitable for CAVs. In this study, we have shown that this is not possible. Moreover, private and public investments must be made at the beginning of the transition period, while the benefits will only become apparent after a longer period of implementation that is accompanied by relevant developments in CAT technologies and successful CAV market penetration. It also remains to be seen which technical standard will underpin the expansion of connectivity: solutions that run adjacent to roads instead of over a wide area would mean large-capacity roads being chosen first for technological upgrades, leading to the exacerbation of existing regional inequalities.
A more effective **management of traffic flows** would allow fuel consumption and thus the production of harmful emissions to be reduced, alongside noise pollution levels.	Traffic management would lead to a stabilization of traffic flows, which would help climate and environmental targets to be met, initially on large-capacity roads; however, it could prove extremely challenging to gain acceptance for the measures, at least initially. Moreover, such reductions alone will not be sufficient to reach the goals set out in various climate targets and strategies; this will require the considerable expansion of post-fossil fuel propulsion systems, new fuels and a greater use of ecomobility options.
CAT can help people **avoid transport**, which opens up the possibility of reclaiming streetscapes and allowing new (urban) activities to take place.	Existing scenarios assume the opposite: empty runs, a broader pool of transport users and the greater convenience offered by such services will lead to an increase in traffic. Only continued rising demand for car and ride sharing will lead to a reduction in the number of registered private cars and thus, potentially, offer scope for parking spaces to be reclaimed. The objective of encouraging users to avoid transport altogether so that public space can be reclaimed can still only be achieved by largely moving away from motorized private transport.
Social groups with limited mobility can (once again) travel independently, thus facilitating their contact with others and helping to improve social cohesion.	This argument must be examined through a more nuanced lens: elderly people with a driving licence will, in the medium term, be able to spend more time using MPT, and even adolescents will be able to drive using CAT in a Long Level 4; however, both increase demand for MPT. Those with a physical or learning disability who require aid when walking as well as when entering or disembarking vehicles are disadvantaged by driverless CAT, and that includes both MPT and public transport services. They thus have to contend with additional barriers. The fact that in future CAT will solely be accessible via apps, networks and clouds will put up further barriers (i.e. create a digital divide).
As **drivers no longer have to control the vehicle**, they can use travel time to engage in other activities; this reduces the burden on the driver and shortens the subjectively perceived journey time.	While travel becomes more comfortable thanks to the possibility of door-to-door transportation, no longer needing to look for a parking space, the ability to use journey time for other activities (work, social media, etc.) and to be provided with uninterrupted infotainment, such convenience and disengagement from the immediate social and spatial environment will do little to boost social cohesion. As the journey time becomes subjectively shorter, the relevance of suburban locations will also increase, especially if they are located close to largely automated transport routes. This will increase the number of longer journeys.

6.2

STRATEGIES FOR SUSTAINABLE TRANSPORT AND URBAN (DISTRICT) DEVELOPMENT

At present, municipal/local-level policy and planning authorities are uncertain about how much influence can and should be exerted on the development of CAT and which mechanisms should be employed to this end. Often the necessary specialist skills, financial resources and time required to play an effective role in shaping the upcoming transformational processes are lacking, and this is especially true at the local level. As our expert impact assessments (see Chap. 3.4) show, in spite of widespread agreement on the urgent need for action, opinion remains divided as to which strategies and resulting action plans should be adopted and how urgently. Although during this early phase of technological development there is, at best, only limited evidence of the possible impacts of CAT, it is vital that key political and planning decisions on the potential applications of CAT be made soon. Above all, it is crucial to address these challenges early on and not to wait until a framework for action has already been firmly established at higher transport policy levels and/or new technologies are already appearing on roads.

Since 2013, the EU has been offering guidance to towns, cities and regions within the bloc to help them develop and evaluate Sustainable Urban Mobility Plans (SUMPs). In June 2019, a SUMP practitioner briefing was released that takes into consideration the challenges (both opportunities and risks) presented by CAT (Backhaus et al. 2019). To ensure progress is made towards the objective of sustainable local and regional development, the primary aim is to empower policy and planning authorities at these levels and facilitate the successful implementation of their strategies. To this end, 12 strategic steps were developed in four phases involving strategy development, implementation and evaluation (see Fig. 6.1.2).

This planning cycle also includes the primary objectives of the AVENUE21 project:

■ Analyse the challenges and opportunities of all modes of transport (Chap. 3.2),

■ Build and jointly assess scenarios (Chaps. 4.1 and 4.2),

■ Select measure packages with stakeholders (Chaps. 7.1, 7.2 and 7.3; see Fig. 6.1.2).

In contrast to the proposals put forward by Backhaus et al. (2019), the AVENUE21 project examines the challenges facing policy and planning authorities during the Long Level 4 transition phase. Taking a sustainability-focused planning concept as a starting point (see also Chap. 3.2), the following section examines the action plans and measures developed around roads as public spaces to formulate integrated mobility development concepts. A key prerequisite here is the setting of clear urban (district) development and mobility targets that make it possible for a transition to sustainable and inclusive urban mobility (DStT 2018) to be initiated and achieved. A shift away from the "car-friendly city" requires a reversal of past developments and decisive action. Here CAT offers both opportunities and risks and presents city-regional administrative and planning bodies with new challenges. As part of a wider mobility shift, it is important to define the requirements and framework conditions for governance and planning processes (adaptive, controlling, restrictive and/or promoting) so that CAVs can contribute to achieving the ambitious aims of sustainable mobility (see Chap. 5).

Figure 6.1.2: **Twelve steps to developing a sustainable mobility plan for towns and cities that considers the implications of connected and automated transport for planning practice (SUMP 2.0)**

PREPARATION AND ANALYSIS

Milestone: analysis of problems and opportunities concluded	**1. Set up working structures**	1.1 Evaluate capacities and resources 1.2 Create interdepartmental core team 1.3 Ensure political and institutional ownership 1.4 Plan stakeholder and citizen involvement
	2. Determine planning framework	2.1 Assess planning requirements and define geographic scope ("functional urban area") 2.2 Link with other planning processes 2.3 Agree timeline and work plan 2.4 Consider getting external support
	3. Analyse mobility situation	3.1 Identify information sources and cooperate with data owners 3.2 Analyse problems and opportunities (all modes)

STRATEGY DEVELOPMENT

Milestone: vision, objectives and targets agreed	**4. Build and jointly assess scenarios**	4.1 Develop scenarios of potential futures 4.2 Discuss scenarios with citizens and stakeholders
	5. Develop vision and objectives with stakeholders	5.1 Agree common vision of mobility and beyond 5.2 Co-create objectives for all modes with stakeholders
	6. Set targets and indicators	6.1 Identify indicators for all objectives 6.2 Agree measurable targets

MEASURE PLANNING

Milestone: Sustainable Urban Mobility Plan adopted	**7. Select measure packages with stakeholders**	7.1 Create and assess long list of measures with stakeholders 7.2 Define integrated measure packages 7.3 Plan measure monitoring and evaluation
	8. Agree actions and responsibilities	8.1 Describe all actions 8.2 Estimate costs and identify funding sources 8.3 Agree priorities, responsibilities and timeline 8.4 Ensure wide political and public support
	9. Prepare for adoption and financing	9.1 Finalize and assure quality of Sustainable Urban Mobility Plan document 9.2 Develop financial plans and agree cost sharing

IMPLEMENTATION AND MONITORING

Milestone: measure implementation evaluated	**10. Manage implementation**	10.1 Coordinate implementation of actions 10.2 Procure goods and services
	11. Monitor, adapt and communicate	11.1 Monitor progress and adapt 11.2 Inform and engage citizens and stakeholders
	12. Review and learn lessons	12.1 Analyse successes and failures 12.2 Share results and lessons learned 12.3 Consider new challenges and solutions

Source: adapted from Backhaus et al. (2019: 11)

6.3

APPROACHES TO PROACTIVELY SHAPING ACTION PLANS, CONCEPTS AND MEASURES FOR CONNECTED AND AUTO- MATED TRANSPORT

Within the context of discussing how CAT can be adapted to meet sustainable development targets through SUMPs, there are several vital details still in need of clarification. No regard has been given to different technological standards (or their hybrid forms in the Long Level 4), various operating design domains (ODD), or different CAV usability options (e.g. automated drivability), and there is no nuanced conception of space, nor are there scenarios in which conceivable future developments are discussed.

Only a nuanced observation of existing policy and planning in towns, cities and regions and of how space is currently used will allow relevant action plans, concepts and measures to be developed within the scope of CAT. We will undertake such an observation in this section, looking specifically at "streetscapes/public space". Vastly different frameworks can be "played out" at this level and relevant concepts and measures can be developed (see Fig. 6.3.1). Applying the framework to a specific neighbourhood not only means examining considerably different road sections, each with their own unique traffic conditions, architectural/physical designs and uses. In addition to the aggregation effect of various road sections, there are also overarching systemic aspects at play, such as integration into the wider urban transport network, organization and allocation of mobility points and hubs for multimodal services, centralized parking options, charging stations in public spaces, local supply of goods provided by delivery services and the organization of the first and last mile as well as the reorganization of certain areas of public space, including approaches to tackling land-use conflicts.

Stepping back and looking at action taken at a higher level, e.g. that impacts links between urban areas and their surrounding regions, represents another observational shift. Here too similar aggregation and concentration phenomena, such as those seen at the neighbourhood level, must be taken into account. Moreover, the city region itself is the level at which sustainable mobility can be safeguarded, as it involves factoring in the nearby areas that are home to commuting workers, students/trainees and shoppers, who rely on (predominantly MPT-based) mobility structures in suburban areas. The city region is also a level that involves vertical integration into the transport and mobility policies of federal states, nation states and the EU. Moreover, city regions also contain rural zones and small-town structures that differ considerably from inner-city districts in terms of their economic and social structures, as well as in terms of where problems and needs arise. A more precise analysis of rural areas is required to understand exactly how this will alter the impact of CAT. The AVENUE21 team will examine this subject in a follow-up project.

We will start by outlining the structure of the relevant levels, action plans and measures (see Fig. 6.3.1).

6.3.1 DESIGNING STREETSCAPES AS PUBLIC SPACE

Status quo:
Public space fulfils a multitude of functions. It not only serves as a space for transport but also as an important stage for public life. People's diverse needs give rise to opposing and competing demands for public space, whether it is somewhere to spend time or talk with friends, or a place where people can walk, cycle, ride a scooter or drive a car. These needs are in constant conflict. Main roads, in particular, are largely characterized by cars and their associated impacts, such as separation effects as well as noise and air pollution, while it is parked cars that usually dominate access roads. There are also growing demands for public space designed to create a sense of identity and to be used in a flexible, multifunctional way, and these changes are necessary for the creation of high-quality public spaces, as they support strategies that promote walking and cycling over motorized transport as well as a compatible traffic handling approach.

Dynamics:
Many of the relics of the car-friendly city, such as inner-city flyovers, town squares used as car parks, large road junctions and traffic distributor roads with ramps, are still a frequent sight in Europe's cities. Building renovations and land requirements necessary for the urban but also economic regeneration of inner cities mean many of these structures already stand ready for use. If

Figure 6.3.1: **Levels, action plans and stances concerning the sustainable implementation of connected and automated vehicles**

	1 **DESIGNING THE STREETSCAPE** **AS A PUBLIC SPACE**	**2** **INTEGRATED MOBILITY** **DEVELOPMENT CONCEPTS**
Action plan A: Integration of CAT into existing transport and mobility systems	**A_1** Plan automated drivability and design the urban-compatible implementation of CAVs	**A_2** Improve multimodal availability and integrate CAT within the neighbourhood
Action plan B: Taking a fair approach to the sharing of public space	**B_1** Reclaim public space	**B_2** Use available land for neighbourhood development
Action plan C: Participation	**C_1** Consider the creation of specific CAT road infrastructure and manage potential land-use conflicts	**C_2** Implement real-life experiments in the neighbourhood
Measure 1:	**1/1** Define a multilevel speed system across various road sections and generally manage speeds (to control traffic flows, emissions)	**1/2** Expand hybrid mobility services as the backbone of urban mobility
Measure 2:	**2/1** Reduce volume of car traffic; boost number of pedestrians and cyclists	**2/2** Change modal split to benefit active mobility and public transport
Measure 3:	**3/1** Improve the quality of streetscapes by designing and ensuring the (fair) shared use of space	**3/2** Define a network of public spaces and active mobility
Measure 4:	**4/1** Designate space for parking, mobility points and charging stations in a way that respects the streetscape's existing characteristics	**4/2** Functionally and architecturally integrate CA mobility hubs as anchor points within the neighbourhood
Measure 5:	**5/1** Organize the first and last mile	**5/2** Develop regional concepts for freight transport, logistics and distribution
Measure 6:	**6/1** Introduce a participative model for the design of public space	**6/2** Jointly develop transdisciplinary visions of "new urban mobility"

Source: AVENUE21

MPT can be reduced to the extent that there is less potential for conflict on main inner-city roads, thus allowing them to be more effectively integrated into a system of public spaces, there will be better opportunities for residents to spend time in public space and for different forms of active mobility (walking, cycling). Improving the quality of public space also increases the quality of living, thus providing greater incentives for investment in necessary renovation work or the construction of buildings. However, such dynamics can also intensify existing processes of gentrification.

The fact that the mobility shift is occurring alongside an ongoing digitalization trend is already giving rise to new stress factors in public space. The growing volume of delivery traffic resulting from the presence of courier, express delivery and parcel services, as well as the rising number of charging stations and micromobility rental services ("free-floating" bike and scooter sharing) are visible examples (see Chap. 3.3). At the same time, lo-

cal authorities' powers to control and influence developments are being diminished. This is partly because many of the new urban means of transport are operated by global companies, but it is also due to the ongoing trend towards privatization and the commodification of public space as well as the growing number of semi-public and private spaces.

So how should we envisage the roads of tomorrow? How will road space currently occupied by parking spaces or lanes be used in the future if such features are no longer needed? Perhaps to create cycle lanes, which are also used by the small electric-powered vehicles (e.g. scooters, delivery bots) that are now entering the market? Or should it be used to create desperately needed space for wider pavements with areas for pedestrians to spend time or relax, for play and exercise zones (for children and adults alike), grass verges or trees? Perhaps local authorities are expecting to generate new sources of income by leasing space to restaurants? And, finally, will these improvements take place away from those areas that are already in the midst of gentrification and seeing its negative effects? All of this illustrates that the redistribution of reclaimed space will not only lead to conflicts between a diverse set of interest groups that will differ widely depending on the location, but that clear objectives also need to be set out that are part of both citywide and neighbourhood-specific urban and district development strategies as well as strategies to develop urban mobility services.

 ACTION PLAN A1:
PLAN AUTOMATED DRIVABILITY AND DESIGN THE
URBAN-COMPATIBLE IMPLEMENTATION OF CAVS

Considerations of streetscapes must take into account a number of factors, including function (e.g. through, connecting or access roads), traffic volume, characteristics of the location (e.g. inner city fringe, 1970s housing estate or a new commercial zone), how the space is used beyond the provision of lanes for traffic, as well as the existing road space design (e.g. boundaries, width and route). The implementation of CAVs should then be planned accordingly. The diverse range of streetscape contexts results in myriad demands being placed upon the application of automated driving systems. If the technology is to be deployed in an environment that is highly complex, for instance, as it is also shared by many pedestrians and cyclists or features several crossings, this increases the demand on CAT. This means that either some parts of the city's road network will remain unsuitable for highly automated driving for many years to come or traffic will repeatedly be brought to a standstill due to "external disruption". Designing an approach that is compatible with the urban area is thus crucial and should play a key role in any CAT impact assessment. A streetscape will be less CAV-compatible in areas that are more sensitive to the use of CAT, where short-distance mobility is more relevant and there is less space available. This is especially true in places where vehicles travel at speeds above 30 km/h. Instead of simply reducing the streetscape to the distance between opposite houses, it is also important to consider how the area at ground level is used.

B1 **ACTION PLAN B1:**
RECLAIM PUBLIC SPACE

Appealing urban public spaces are those that provide areas where people can relax or exercise but that are designed to offer the greatest possible access to weaker road users, e.g. cyclists, pedestrians and those with limited mobility. The unusual way in which humans and machines communicate as part of CAT, i.e. where no eye contact is required, could cause additional insecurities to arise in public space, thus damaging the appeal of the

space itself. New CAT use cases, such as automated delivery robots on wheels that use pavements, or e-scooters, are already creating anxiety and disruption as well as giving rise to new conflicts over the use of public space.

If a large number of conventional, privately owned cars are replaced by more "intelligent" models, there is concern that the increased use of CAT will result in the continued unchallenged dominance of MPT. As a result, the existing strains that can be seen in urban areas will not only remain, but might even become further exacerbated (i.e. more journeys). There is reason to fear that the much-needed mobility transformation will be hampered as a result and that a city designed around CAT will be promoted, which would be a step backwards to the logics that are being criticized today.

However, CAT also offers opportunities: fewer cars in towns and cities means fewer parking spaces in public areas. Yet this can only be achieved when private vehicles are replaced by car- or ride-sharing vehicles. At present, it remains impossible to predict how successful this could be given the socially selective levels of acceptance, regional differences in service provision and the technological and economic limitations in Level 4 (an automatic reduction might only be possible on selected sections of road). In spite of the considerable differences in demand, recent studies suggest that one car-sharing vehicle could replace up to ten private cars (see Chap. 3.3). Furthermore, a systematic shift in urban mobility will not only free up parking spaces, it could also potentially impact additional functions associated with motor vehicle use (e.g. garages, car dealerships).

C1 **ACTION PLAN C1:**
CONSIDER THE CREATION OF SPECIFIC CAT
ROAD INFRASTRUCTURE AND MANAGE POTEN-
TIAL LAND-USE CONFLICTS

CA, electric and micromobility vehicles require additional space to travel (i.e. lanes) and for the relevant infrastructure (mobility points, parking spaces, charging stations, sensors, etc.). These spaces must be created within existing open areas on and around roads, which are often limited and already heavily used. Mobility points would either need to be integrated into existing buildings, could be placed in areas previously used by private cars or car-related services, or require separate structures with appropriate entry and exit slip lanes. The desired electromobility expansion will encounter difficulties on roads surrounding perimeter block developments or restricted by high-rise apartment buildings as it will be challenging to find and design suitable locations for charging stations. Highly automated vehicles must be designed so that they can autonomously connect to and disconnect from charging stations, so the latter take up little space and the former can be charged in little time.

Even if it is possible to gain access to certain parts of the streetscape (traffic lanes and lay-bys in particular) and spaces reserved for cars/car-related services, it remains to be seen how these spaces can be used in a way that is city-friendly, that improves the atmosphere or that contributes to improved active mobility. Depending on the situation on specific sections of road, it is necessary not only to make preliminary transport and urban planning decisions but to do so with the adequate inclusion of additional stakeholders (business and homeowners) and those living in the relevant commuter area. This is unlikely to proceed without conflict, especially if during the transition phase less space is "reclaimed" than initially expected.

These action plans give rise to the following six measures for public streetscapes:

 Measure 1/1: Define a multilevel speed limit system across various road sections and generally manage speeds

One objective of improving the quality of urban public space is to make it easier for several road users to travel simultaneously as well as for people to comfortably share the same space. CAVs that travel at slower speeds help improve the level of safety for other road users in shared areas. Travelling more slowly will help reduce the demands placed on the technology, but it will also lower the "transport capacity" of the existing infrastructure.

Making CAT compatible with urban areas would thus mean setting temporary speed limits along certain sections of road (walking pace, 30 km/h zones, etc.). Vehicle automation and a connected traffic management system mean speed limits would be observed, which will test the patience of some road users.

 Measure 2/1: Reduce volume of car traffic; boost number of pedestrians and cyclists

How compatible transport modes are with an urban area and how well different users can effectively and fairly share public space are determined not only by speed but by the volume of car traffic. Recent projections, in particular, suggest that automation will lead to the age range of passengers increasing, and that convenience and travellers' spontaneity will result in more trips (see Chap. 4.3). To rectify the errors made by the car-friendly city and to incorporate multifunctional major inner-city roads into an urban context in a way that is compatible with other objectives, it will be necessary to reduce the number of cars on the roads. Furthermore, an optimal transport mix with a higher percentage of routes completed on foot or by bike, and transport links that are appropriately developed to meet this aim will be required to ensure safe and suitable shared road space for all users.

 Measure 3/1: Improve the quality of streetscapes by designing and ensuring the (fair) shared use of space

A substantial force driving the mobility transformation is the call to create pleasant areas within the streetscape where weaker road users (such as pedestrians, cyclists and those with restricted mobility) are able to relax and exercise safely. This will require a rethink of how road spaces are shared, and greater design emphasis should be placed on making spaces more pleasant and encouraging local mobility. Urban planning criteria stipulate the application of four principles here: (1) reduce the number of car parking zones in public spaces; reorganize waiting, loading and deliveries at designated drop-off and pick-up zones (see measures 4/1 and 5.1), (2) minimize lane width thanks to CAVs' adaptive cruise control safety feature, allowing proportionality, (3) create more space for pedestrians and cyclists as well as for adjacent green and open spaces that improve the overall urban atmosphere, and (4) reduce various mobility restrictions.

 Measure 4/1: Designate space for parking, mobility points and charging stations in a way that respects the streetscape's existing characteristics

Until now we have only been offered highly idealized visions of how the additional demands of connected and automated MPT, shuttle buses and ride-sharing vehicles, electromobility, new micromobility vehicles, delivery services and, finally, the growing demand for active mobility will bring about changes to existing traffic lanes and infrastructures. Plans detailing ideal scenarios envisage streetscapes that are between 60 and 80 metres wide (NACTO 2017) – such dimensions are rare in European cities – and an almost universal separation of lanes for different vehicle types. This risks creating new "barriers" and new, more intense conflicts over how the streetscape is used as part of the "new urban mobility".

In transit areas, and at industrial/production sites as well as supra-regional transport hubs, where roadside activities are usually less sensitive and vulnerable, it could be possible to implement CAT in the relatively near future. Measures to adapt road infrastructure to the needs of connected and automated transport must, where necessary, be meticulously planned and implemented, and their potential to cause disruption appropriately evaluated. Transport researchers and planners who are part of district and local development teams should be tasked with deciding how these new, greater demands on the streetscape can be taken into account and integrated in an effective way. How and when further stakeholders and, if appropriate, citizens should be included in the decision-making process are questions that should be reserved solely for very small spaces and certain sections of road (see measure 6/1).

A possible solution for inner-city neighbourhoods is the division of space by specialization, i.e. dividing space into even more sections based on which type of activity or mobility is available and where (similar to systems already in place for one-way streets, bus and taxi lanes, shared spaces or "play streets"). To implement this measure in an effective way, the road infrastructure will need to be adapted (e.g. via the introduction of corridors) as part of a higher-level strategy that takes into account network hierarchies and other, potentially untapped functions located alongside streetscapes.

⊞ Measure 5/1: Organize the first and last mile

How the first and last mile is organized for the movement of both people and goods will become a core task over the course of a Long Level 4 during which only a portion of the transport network will be ready for CAT rollout. This will apply within urban areas as well as along public transport routes that will need to be maintained and further expanded as part of the mobility transformation. If CAVs were to be used for last-mile journeys that are predominantly completed on foot, this would be counterproductive to achieving the stated aim of boosting ecomobility and improving public health. This is something that needs to be considered during streetscape design. In outlying districts and areas not intersected by transport routes, the targeted use of CA shuttles could considerably increase public transport accessibility and, if efforts to reduce MPT are successful, lessen some of the stresses currently burdening streetscapes. Specialized CAVs could thus become a temporary solution to help gradually improve the atmosphere of heavily used roads and boost the appeal of active mobility. The same may be possible for goods transport, but here the key difference is that, at present, there are hardly any publicly run services.

⊞ Measure 6/1: Introduce a participative model for the design of public space

In the discussion concerning how public space should be divided – space that CAT could help reclaim – there is often little mention of how decisions on the space's future use and possible redevelopment should be made. The need to establish a new, just distribution of public space should be addressed as part of a sustainable transport and mobility plan. Only within the context of these stipulations will it be possible to discuss, and potentially also decide upon, functional and design alternatives as part of participatory processes. Against the backdrop of diverse societal interests and a widespread shortage of space in inner-city neighbourhoods, when it comes to deciding how public space should be repurposed, there are, unsurprisingly, a wide range of opinions. These processes should thus be designed in such a way that neither solely economic nor another single set

of interests is able to prevail, but that active mobility, a functional mix, and the aesthetic appeal of the space itself are enhanced to benefit as many social groups as possible while reinforcing the idea of sufficiency. This would dramatically improve the chances of a successful mobility transformation.

6.3.2 INTEGRATED MOBILITY DEVELOPMENT CONCEPTS

Status quo:
Neighbourhoods and urban regions encompass subspaces with a range of functionalities, sites, residential population compositions, mixed uses, etc. It is thus expected that CAT will have a broad range of impacts, but in spite of this knowledge, there has been almost no nuanced analysis of the effects. As the analysis of streetscapes and public spaces has shown, CAVs will be especially hard to implement in urban transport settings where different road users occupy the same space. By the same token, it can be expected that there will be neighbourhoods, sites and locations that not only make way for CAT but can in fact benefit from this technology if it is adapted in a certain way. If the focus is on meeting the objectives of the mobility transformation and sustainable city and district development, it is vital to ask how CAT can be implemented to safeguard or even improve the functionality of different neighbourhoods.

Dynamics:
The development pathway followed by subspaces in city regions depends on their location, facilities and accessibility, the level of economic development and the composition of the residential population. These trajectories differ vastly, which results, on the one hand, in outmigration, disinvestment, abandoned property and underused public space and, on the other, in growth pressure, which subsequently entails competing demands for space. CAT can only marginally counteract this trend as automated drivability will tend to be high in areas where streetscapes are largely empty and homogeneous. Conversely, CAT can lead to considerable disruption and destabilization in "lively" and densely used areas, if permission is even granted for CA vehicles in such zones during Level 4 (see "Risks to transport safety in the Long Level 4" in Figure 6.3.1).

A2 ACTION PLAN A2:
IMPROVE MULTIMODAL AVAILABILITY AND INTEGRATE CAT WITHIN THE NEIGHBOURHOOD

The quality of a neighbourhood is heavily determined by the level of available services, accessibility, multimodal transport services and the quality of public space. Thus any implementation of CAT should take into consideration the relevant advantages and disadvantages of various use cases and do so without favouring individual transport means. For instance, drop-off and pick-up zones in outskirts and peripheral areas for a CA service that has been integrated into the public transport network – similar to those used by local suppliers – could be reached on foot in less than five minutes. To achieve this, mobility hubs will become increasingly important as elemental transfer points for seamless intermodal movement at the interface between neighbourhoods, the city as a whole and the wider city region and in all their relevant subspaces. One task that planning teams will have to tackle is to weigh up the choice between fixed-schedule routes with defined stops and an on-demand door-to-door service with greater spatial flexibility. The high capacities offered by traditional scheduled transport services, such as underground lines, commuter trains, trams and buses (which could also employ automated operation) are maintained, thus allowing transport within the city region to be consolidated, even with CA and its associated requirements remaining key and growing in importance in light of the mobility transformation.

The location of mobility hubs and their role within the city region's transport system and its subspaces will partly be determined by the potential for demand to increase (due to the number of residents and employment opportunities), land availability and proximity to traditional public transport services. At certain sites, they will offer opportunities to improve accessibility and to encourage the formation of centres.

 ACTION PLAN B2:
USE AVAILABLE LAND FOR NEIGHBOURHOOD DEVELOPMENT

In addition to the previously mentioned possibilities of new public space becoming available through the removal of parking spaces, the CAT transformation may also free up other space in a neighbourhood. Precisely which areas will depend heavily on the structure of the future mobility market (see Chap. 5). Structural changes will potentially impact sectors associated with the car industry, such as parking garages, parking spaces, petrol stations, car washes, car workshops and car dealerships, that may gradually change. The buildings or spaces they currently occupy could also be freed up for other uses. Once integrated within a strong infrastructure, these spaces will become potential areas for sustainable brownfield regeneration projects based on urban principles. To develop these spaces in accordance with existing objectives, integrated concepts for district and neighbourhood development will be required that effectively combine issues and areas of activity that are relevant to the development in question, such as those concerning settlement and open space development, infrastructure and mobility. This will give core urban areas the opportunity to improve the quality of their public spaces and green areas, and for monofunctional spaces,

this will boost urban diversity by expanding and concentrating the ways in which the space is used.

However, it should be noted that new micromobility vehicles require new mobility hubs as well as the relevant electromobility infrastructure (e.g. charging stations) and CAVs also require space. This will necessitate the large-scale freeing up of areas within public space.

 ACTION PLAN C2:
IMPLEMENT REAL-LIFE EXPERIMENTS IN THE NEIGHBOURHOOD

To be able to prepare for the future against a backdrop of uncertainty and diverging objectives, reflexive approaches were created within urban (regional) development strategies ("targeted incrementalism") that are now the subject of broad discussion both concerning their theoretical approach and their practical implications in urban planning (see Chap. 4.7). The challenges involved in introducing CAT to historical European cities demand a gradual and reflexive approach. As analysis of our ideal future scenarios shows, however, the decision on how CAT should be implemented is crucial for determining whether future mobility will pursue the existing objectives of a sustainable city-regional development or whether the direct and indirect effects will be counterproductive to these aims. One element that plays a decisive role for the development of CAT is the fact that most of the ongoing real-life experiments involving the technology follow a market-driven approach (see Chap. 5.3). Whenever towns and cities play an active role in such trials, it is often purely to market the town/city or district, with civil society only rarely having a defined role or being given any real opportunities to shape policy.

Within the context of these real-life trials and within a broader understanding of testing, including living labs, alternative mobility concepts, how they are embedded within the neighbourhood and their impacts need to be tested and analysed. Moreover, it is necessary to intensify a dialogue between towns/cities and city regions as part of urban and mobility planning and to incorporate citizens' everyday and practical knowledge into the innovation process. The biggest challenge most probably lies in encouraging all stakeholders to consider the motivation for their current policies or actions. Neighbourhoods and city regions are important drivers of innovation and transformation processes, and are thus vital frames of reference.

This approach appears to be particularly relevant for CAT: firstly, those who are interested can experience and try out previously unknown new mobility services. This allows potential reservations (e.g. how willing an individual is to be in close proximity to other passengers when ride sharing) to be recognized and collective everyday experiences during early development phases to be fed back into the innovation cycle (Chan/Shaheen 2012; see also Fig. 4.2.4). Secondly, it enables an understanding of how new mobility services will interact with mobility and consumer behaviour (the effects of which are still largely unknown) to better gauge the transformative potential as well as the rebound effects. Thirdly, collective learning processes can be initiated that provide information on how social mobility practices are changing, how technology can be integrated into public space and how processes can be designed to safeguard access to new forms of mobility.

 Measure 1/2: Expand hybrid mobility services as the backbone of urban mobility

The dispersed means of transport within a city region require the expansion of a public transport network that is fast, convenient, secure and effective in subspaces where it was previously not economically viable to provide services. Driverless automated public transport offers economic benefits, e.g. allowing more frequent services, longer operating hours and additional (flexible) routes, especially when using smaller vehicles. Creating a needs-based transport service that involves automated shuttle buses of different sizes and serves areas with low population density both with and without permanent stops will improve convenience. On the other hand, CAT (especially if privately owned CAVs prevail) could strengthen and help support dispersed, monofunctional settlement structures at the edges of towns and cities and in areas between transport routes. Here the task for urban planners is to combine new hybrid mobility services with a large-capacity public transport network. Public transport networks must be systematically expanded particularly in city-regional areas where new high-density sites or settlement developments are being built or planned.

Linking and integrating these locations with and into strong public transport systems and boosting their functionality is vital to reducing the traffic burdens on inner towns and cities. Throughout it must be remembered that a huge amount of work must be done in city-regional public transport and also in terms of bicycle mobility, especially along tangential links, to modernize these networks and this need for action could grow further.

 Measure 2/2: Change modal split to benefit active mobility and public transport

During the course of the much-needed mobility transformation, it will be important to gradually shift away from MPT and towards active mobility (walking, cycling) and eco-friendly public transport, collectively referred to as "ecomobility". This can be achieved by imposing bans and requirements (at present less frequently used by policymakers), offering an improved service (cycle lanes, recreational spaces in public areas, stops that are serviced longer and more frequently, punctuality, modern public transport vehicles) and through adapted mobility behaviour (see policy- and civil society-driven scenarios in Chaps. 5.4 and 5.5). Above all, the provision of information about (improved) services, completing certain journeys without a private car and mobility points that are well integrated into the urban landscape are of vital importance. Within this context, non-automated and automated car, bike and, above all, ride sharing are just as important as flexible and hybrid forms of public transport. It remains to be seen how beneficial new mobility services brought about by the mobility transformation will be (e-scooters being a prime example). If these new means of transport are predominantly used as a substitute for walking, the effect will by no means be positive.

As soon as they are rolled out on parts of the road network, above all on motorways (low complexity) or special zones (low speed; see Fig. 4.1.7), connected and automated vehicles change the overall transport situation. Sharing and public transport services can then be expanded and made more affordable, thus boosting demand, which will ultimately facilitate the mobility transformation. During the transition phase, the primary expectation is for increased pressure on further urban sprawl brought about by highly convenient transport options that will offer serious competition, not just to existing public transport services (e.g. caused by a loss of their USP). Moreover, without appropriate intervention, social cohesion could be undermined in several ways if more aspects of journeys are delegated to machines. Those considering realistic alternatives to MPT today should not overlook the fact that highly or fully automated private cars are likely to be extremely comfortable and highly appealing.

 Measure 3/2: Define a network of public spaces and active mobility

Public space is not just an isolated square or a park: it is a network connected by roads and paths and is used as such. Any transport plan that aims to support the mobility shift should not focus solely on the flow of transport as its key issue. Quite the opposite: the city region of the future should be considered and developed as a public space network. Plans for new housing developments surely offer a far easier opportunity to implement this demand as opposed to upgrading historical buildings in Europe's cities, but even here resistance must be expected.

In existing neighbourhoods close to the inner city, a network of public spaces should be made possible through an increased differentiation of individual street sections, which must be based on their various suitability for the use of CAVs. However, this will mean highly complex negotiation processes with ground-level property users, as well as homeowners and residents, albeit within a set framework that has been carefully and objectively considered by planners. It is essential to avoid a general increase in automated drivability becoming a transport policy priority. Should this occur, earlier planning approaches based on a car-friendly city would be repeated with zero lessons learned.

 Measure 4/2: Functionally and architecturally integrate CA mobility hubs as anchor points within the neighbourhood

Mobility hubs are appropriate anchor points for neighbourhood development and should be designed as such. For this to happen, suitable land with good public transport accessibility needs to be secured in good time. CA mobility hubs can provide impetus to improve the overall quality of an urban area, especially in neighbourhoods without a real centre. Making mobility hubs and their immediate surroundings more aesthetically pleasing offers an opportunity for more enticing urban features, which could especially benefit housing estates built between the 1960s and 1980s, which were planned and constructed under the mass motorization model, and business and production sites. Moreover, the mix of mobility services, especially given the amount of land required for each, should be considered. The integration and combination with other facilities, such as parcel drop-off points, kiosks, bakeries, restaurants and market stalls, increase points of contact and thus ensure there is the necessary demand for all services. Such measures to improve neighbourhoods can also help increase acceptance of CAT.

 Measure 5/2: Develop regional concepts for freight transport, logistics and distribution

The transformation processes heralded by CAT will also significantly impact the supply networks of city regions. Against the backdrop of international competition, pressure to keep costs low and the constantly growing demands of customers for fast, scheduled deliveries of goods, the pressure to develop more efficient, multimodal logistics and distribution concepts is steadily rising. A Long Level 4 will give connected and automated vehicles the opportunity to demonstrate their strengths, especially in logistics and distribution. This will ultimately also lead to a faster transition of vehicle fleets to CA models, as well as to the development of new vehicles and delivery services. The coordinated hierarchy of logistics and transport sites will (once again) make the development of new sites necessary. This is particularly the case for – primarily last-mile – shuttle services, which will need to be adapted to specific urban areas.

Cities and local authorities thus cannot afford to take a "wait and see" approach but instead need to become actively involved in these highly dynamic development processes. This includes a strategic location and land use management plan that is part of a wider coordinated system of regional and urban logistics sites. Urban services, especially last-mile solutions, must be designed to be sustainable and city-friendly – with regard to the design of new city hubs, this means the use of city-friendly vehicles and/or the licensing of delivery services at the urban level.

 Measure 6/2: Jointly develop transdisciplinary visions of "new urban mobility"

How digitalization will change city regions and "new mobility" in years to come must become the subject of wide-ranging discussions within civil society and among policymakers, planners, researchers and businesses. At present, possible visions of the future shown in the media are influenced by those tech companies leading the digitalization revolution: the impacts of technologies on city regions and new mobility are only seen from a single viewpoint. By comparison, structures enabling transdisciplinary "living labs" could encourage societal processes involving negotiation between and production involving a variety of actors. Generally speaking, living labs that are based on a collective vision of the future shared by all actors make it possible to pinpoint specific challenges faced by city regions, to reach a consensus regarding objectives, lay bare disagreements, and translate goals into concrete action. For this to be achieved, participatory processes need to be designed and implemented.

6.3.3 THE GROWING IMPORTANCE OF THE CITY REGION

Strong connections and transport links between a city and its surrounding area are what makes city regions ideal case study candidates for settlement development: analysis of this specific type of settlement enables effective strategical decision-making for sustainable development. It must also be assumed that the relevance of city regions for transport and settlement policy will only grow in years to come, making close cooperation between highly unequal partners inevitable. In terms of horizontal integration, this means cooperation between various regional authorities, but it is vital that no one group is able to pursue their individual interests unchallenged. Ensuring a fair approach is often extremely difficult, especially when choosing sites for new office blocks or single-family homes.

The inherent challenges of integrating the specific interests of specialist policymakers and planners into spatial planning have a unique impact at the city-regional level. Unlike neighbourhoods or small- and medium-sized towns, city regions play a vastly different role in terms of vertical integration as they represent a key strategic level.

For the implementation of CAT, it will be vital to integrate transport and mobility planning into settlement development at the city-regional level. To keep in line with the vision of a mobility transformation, it will be necessary for most settlement development plans to be based along higher-level public transport routes or on developing existing stock. At the city-regional level, CA-based public transport and intermodal mobility hubs must be planned and created within a hierarchical network; the necessary infrastructure must also be designed, financed and operated. For this to happen, integrated mobility services need to be developed and used, and, ultimately, access to any generated data must be safeguarded.

One major factor is the issue that city regions also encompass rural areas, small towns and villages in which facilities and accessibility and, subsequently, transport services and mobility styles differ significantly from those in cities. There is currently little to no accurate information on the pros and cons of CAT in rural areas. For instance, if the development of rural areas and equivalent living standards are to be ensured, it is vital that CAT be supported by a nationwide 5G network. With such a system in place, it will be possible to employ the technology in rural regions: indeed, as their transport systems are less complex, they would be best suited as CAT testbeds. It would be possible to create wide-scale automated drivability more rapidly in rural zones, which for some areas would bring new opportunities for stability and growth.

Next year, the AVENUE21 team will turn their attention to rural spaces while also devoting more attention to the strategical importance of city regions in the development and implementation of CAT

6.3.4   OUTLOOK: THE RELEVANCE OF DATA

The digital transformation will have a profound influence on all areas of everyday urban life. In future, data-based processes will dramatically change how we shop, how mobility is designed, how energy is generated and consumed, how goods are produced and services delivered, and how we communicate with one another. New business and civil society actors are entering the scene, which will undoubtedly have an impact on traditional duties and roles within urban planning.

The availability of data will take on even greater relevance as such information can be almost limitlessly "produced" by every individual, and algorithms developed by artificial intelligence will open up new opportunities for companies. Through analysis, these data can be harnessed to facilitate efforts to tackle societal challenges, but they also create new needs and dependencies. This is particularly relevant to the transport system and mobility behaviour. For cities and city regions, the possibility of being able to bring added efficiency and safety to transport management systems in the future means creating and operating data collection, storage and processing infrastructure designed for the common good.

If cities and city regions take over responsibility for transport management and information or wish to take on private companies and compete as a mobility service provider, this will also create a need for additional competencies as well as for the regulation of access to locally generated data. Cities and city regions alone will struggle to meet this challenge as central legislation (e.g. concerning cybersecurity or data protection) is the responsibility of nation states or the European Union. This illustrates how great the need is for cooperation at every state-controlled level. The question of how to handle these data thus requires a high level of data competency in planning and administration departments, which is so far lacking in most cases.

It is foreseeable that the digital transformation will, in part, lead to the complexity of planning processes increasing considerably. Knowledge across disciplines, cooperation across departments and institutions, and inter- and transdisciplinary action will become even more important. Cities should not wait for change to come; they should make sure they are ready now and take the necessary cooperative steps.

6.3.5   OPPORTUNITIES AND APPROACHES TO LOCAL AND REGIONAL GOVERNANCE

The scenarios presented in this study have shown that the stance taken by the various groups of actors involved and their various levels of influence will determine the framework for future settlement and transport policy developments. The shift towards a new mobility model has resulted in a multitude of international providers jostling for position in expanding mobility markets and functional spaces. The structure of local and regional governance is of central importance. In light of the major change that awaits the mobility system, it is also vital that existing institutions' capacity for action be subjected to scrutiny. Where necessary, new alliances need to be formed and efforts must be made to accelerate and safeguard opportunities for participation. This is a task that can and must be attended to now.

The developments that have taken place over recent years – collectively referred to, discussed and marketed as the "smart city" – are now starting to highlight the very problems that should be avoided. The technology-driven globalization of local government services and organizational processes has partly resulted in the complexities and contradictions that ultimately make up a city's DNA being once again subjected to the principal of efficiency. The smart city has seen the return of a modernist reductionism that is now trying to infiltrate private spheres of life that were previously off-limits to any such form of control. Upon closer inspection, most invitations to participate in fact turn out to mean nothing more than the granting of third-party access to user data. Discussions concerning how connected and automated mobility will be developed must undoubtedly involve as many areas of society as possible and these groups should also be given an active role in shaping its design. This will require robust local networks that represent a highly diverse range of concerns.

Figure 6.3.2: **Locating action plans within an urban development project in Vienna's Erdberg neighbourhood**

CONSIDER THE CREATION OF SPECIFIC CAT ROAD INFRASTRUCTURE

USE AVAILABLE LAND FOR NEIGHBOURHOOD DEVELOPMENT

IMPROVE MULTIMODAL AVAILABILITY

Illustration: Alexander Diem

RESEARCH TEAM

AVENUE21

7

THE AVENUE21 RESEARCH TEAM

IAN BANERJEE

"What kind of cities do we want to live in? They can be many things: enterprising, equitable, sustainable, smart, creative. I am interested in how a political discourse can be conducted in such a way that CAT is integrated into an overriding paradigm that appropriately deals with social and ecological challenges."

MARTIN BERGER

"How will automated driving change our everyday lives? We need to recognize the many opportunities and risks, effects and rebounds if we are to understand who the winners and losers will be. So much remains uncertain: the future is unknown and no one can precisely predict what will happen. Yet it is important that we consider the shape we would like our urban lives to take when highly or fully automated driving eventually becomes a reality. Will shared vehicles displace private cars? Will certain localities become more or less attractive and thus impact people's choice of where they live, work, shop and relax?"

EMILIA M. BRUCK

"Technological innovations such as vehicle automation not only influence how humans, information and goods travel; ultimately, they also change our urban reality and our consciousness. These technologies impact all aspects of our urban lives, from our employment conditions to our social relationships, patterns of consumption and environmental design. Urban planning strategies should thus not only consider the technological application or use of this new medium but scrutinize the underlying principles and logic that could lead to a complete restructuring of space, society and the economy."

Ian Banerjee is a senior researcher at the TU Wien's Centre of Sociology.

Martin Berger is a professor and head of the TU Wien's Transportation System Planning Research Unit.

Emilia M. Bruck is currently completing her dissertation at the TU Wien's Centre of Local Planning.

JENS S. DANGSCHAT

"Connected and automated mobility is a key element of the digitalization process and the Internet of Things. The most important changes affect not only road traffic but our everyday realities (mobility, communication) and our professional lives (structural unemployment, new professions, greater spatial and temporal flexibility). These factors must be taken into consideration. We are facing a fundamental social shift: we will see existing inequalities exacerbated and new ways of living take shape. A chasm will become entrenched between the winners and losers in regions, sectors and social groups."

ALEXANDER DIEM

"Connected and automated mobility will change building typologies. Circulation space design will have to adapt to a new, complex mobility behaviour and change the appearance of buildings and spatial structures accordingly. As was the case during previous phases of mass mobilization, we will increasingly see the emergence of brand-new types of buildings. Architects and planners will need to meet the challenge of developing an interior circulation design that is suited to a changing mobility system."

JULIA DORNER

"In my view, the most fascinating aspect of connected and automated vehicles is their ability to influence life within a society. In recent decades, we have seen new technologies become increasingly embedded in our everyday lives and how this has, in turn, affected our interactions with one another. I hope that this project will offer an insight into what shape future transport systems might take and allow a greater understanding of how they could influence our future societies."

Jens S. Dangschat is Emeritus Professor of Sociology at the TU Wien.

Alexander Diem is a freelance architect and research assistant at the TU Wien's Architectural Theory Research Unit.

Julia Dorner is a research assistant at the TU Wien's Transportation System Planning Research Unit.

JONATHAN FETKA

"Of course, it is impossible to predict the future, but it is important to explore basic possibilities by developing certain scenarios. A holistic consideration of the possible impacts of new technologies is crucial in order to understand their implications for the urban environment and our society. When it comes to the evolution of mobility, how must planners interpret and reflect on the available information and adjust their approaches to avoid the mistakes of the past? How can we integrate CAT into our ongoing efforts and strategies to resolve the negative effects of previous car-based planning?"

CHARLOTTE HELLER

"Connected and automated mobility will not only change the modes of transport used in our towns and cities, as well as for national and international journeys, it will also affect our future spatial and infrastructural needs. The emergence of new mobility systems and services will necessitate new solutions in the field of cross-border planning and management in order to tackle changing transport habits and traffic volumes."

MATHIAS MITTEREGGER

"In the early stages of a technology's development, much remains uncertain. It is during these years that the new phenomenon becomes apparent; some possibilities seem promising while others appear alarming. These typically eclectic but productive times are of major interest, even those now in the past. The question we should be asking is: what can we learn from history?"

Jonathan Fetka is a project assistant for the future.lab Research Center at the TU Wien's Faculty of Architecture and Planning.

Charlotte Heller is a project assistant at the TU Wien's Centre of Local Planning.

Mathias Mitteregger heads the AVE-NUE21 project at future.lab at the TU Wien's Faculty of Architecture and Planning.

RUDOLF SCHEUVENS

"The rapid pace of automation and digitalization permeates and shapes the city, its urban fabric and its infrastructure. It gives rise to a plethora of questions that must be explored. The most important being: into which future visions of urbanity and urban life can and must these developments be integrated?"

AGGELOS SOTEROPOULOS

"As was the case with previous technological innovations in the field of mobility, connected and automated vehicles will impact the future of mobility in European cities and their settlement structures. In my view, a comprehensive study of the potential impacts and interplay between different elements is particularly important as history shows us that transport is not only subject to socio-technological restrictions but is just as heavily influenced by policy and planning choices. Actors in the latter domains need an optimal foundation on which to base their decisions."

ANDREA STICKLER

"Political hopes and economic interests shape the dominant narratives on connected and automated vehicles and thus also influence local applications. While most questions concern the 'what' (applications and services) and the 'when' (timescale for CAT's implementation), I am mainly interested in the 'where' and the 'how'. This perspective can lead to new insights about necessary regulatory considerations and approaches related to CAT."

Rudolf Scheuvens is a professor of local planning and Dean of the Faculty of Architecture and Planning, TU Wien.

Aggelos Soteropoulos is writing his dissertation at the TU Wien's Transportation System Planning Research Unit.

Andrea Stickler completed her dissertation at the TU Wien's Centre of Sociology.

BIBLIOGRAPHY

8

BIBLIOGRAPHY

Aapaoja, A., J. Eckhardt and L. Nykänen 2017. *Business models for MaaS*. 1st International Conference on Mobility as a Service, 28–29 November 2017. Tampere: Tampere University of Technology.

Abbott, J. 2012. "Planning as Managing Uncertainty: Making the 1996 Livable Region Strategic Plan for Greater Vancouver", in *Planning Practice and Research* (27) 5, 571–593.

Ahmadpour, N., M. Kühne, J.-M. Robert and P. Vink 2016. "Attitudes towards personal and shared space during the flight", in *Work 54*, 981–987.

Akademie für Raumforschung Landesplanung (ed.) 1978. *Beiträge zu Problemen der Suburbanisierung 2*, research and conference reports, vol. 125. Hannover: Hermann Schroedel.

Alberts, V., M. Dirnwöber, F. Kressler, J. Liebermann and K. Stupnik 2016. *Vom Intelligenten Verkehrssystem zum Integrierten Verkehrssystem. Einfach, vernetzt, digital und nachhaltig.* ITS-Austria, Forschung – Technologie – Innovation. Handlungsoptionen 2020+. Vienna. www.smart-mobility.at/fileadmin/media_data/FTI_Roadmap_v5.0.0_final.pdf (6/3/2019).

Albrechts, L., A. Balducci and J. Hillier (eds.) 2016. *Situated Practices of Strategic Planning: An International Perspective.* Milton Park: Routledge.

Alessandrini, A., A. Campagna, P. Delle Site, F. Flilippi and L. Persia 2015. "Automated Vehicles and the Rethinking of Mobility and Cities", in *Transportation Research Procedia 5*, 145–160.

Alkim, T. 2018. *The ODD framework.* Presentation at the Automated Vehicle Symposium, 11/7/2018, San Francisco. Breakout session no. 34.

Allen, A., A. Lampis and M. Swilling (eds.) 2016. *Untamed Urbansm.* London/New York: Routledge.

Allmendinger, P., and M. Tewdwr-Jones (eds.) 2002. *Planning Future: New Directions for Planning Theory.* London/New York: Routledge.

Alonso-González, M., N. van Oort, O. Cats and S. Hoogendoorn 2017. "Urban Demand Responsive Transport in the Mobility as a Service Ecosystem: Its Role and Potential Market Share". Lecture at the International Conference Series on Competition and Ownership in Land Passenger Transport (Thredbo 15), Stockholm.

Alpkokin, P. 2012. "Historical and critical review of spatial and transport planning in the Netherlands", in *Land Use Policy 29*, 536–547.

Altenburg, S., H.-P. Kienzler and A. Auf der Maur 2018. *Einführung von Automatisierungsfunktionen in der Pkw-Flotte. Auswirkungen auf Bestand und Sicherheit.* Basel: Prognos AG.

Altrock, U., R. Kunze, E. Pahl-Weber and D. Schubert (eds.) 2009. *Megacities und Stadterneuerung.* Jahrbuch Stadterneuerung 2009. Berlin: Universitätsverlag der TU Berlin.

Amano H., and T. Uchimura 2016. "A National Project in Japan: Innovation of Automated Driving for Universal Services". In Meyer/Beiker (eds.), 15.

Amano H., and T. Uchimura 2018. "Latest Development in SIP-Adus and Related Activities in Japan". In Meyer/Beiker (eds.), 15–24.

Anderson, J. M., N. Kalra, K. D. Standley, P. Sorensen, C. Samaras and O. A. Oluwatola 2016. *Autonomous Vehicle Technology: A Guide for Policymakers. Santa* Monica: RAND Corporation.

Anderson, J. R., and R. Thompson 1989. "Use of analogy in a production system architecture", in Vosniadou/Ortony (eds.), 267–297.

Ansell, C., and J. Torfing 2016. "Introduction: theories of governance", in Ansell/Torfing (eds.), 1–17.

Ansell, C., and J. Torfing (eds.) 2016. *Handbook on theories of governance.* London: Edward Elgar Publishing.

Aschauer, D. A. 1989. "Is public expenditure productive?", in *Journal of Monetary Economics (23)* 2, 177–200.

AustriaTech 2018. *Elektromobilität in Österreich 2017/18 – Highlights.* www.bmvit.gv.at/verkehr/elektromobilitaet/downloads/emobil_2017_highlights_ua.pdf (6/3/2019).

Axhausen, K. W. 2016. "Autonome Fahrzeuge – Erste Überlegungen". Presentation in the context of the Sommerakademie der Studienstiftung (September 2016). Magliaso: Netzwerk Stadt und Landschaft.

Bach, S. 2013. "Einkommens- und Vermögensverteilung in Deutschland", in *Aus Politik und Zeitgeschichte 10–11/2013.* www.bpb.de/apuz/155705/einkommens-und-vermoegensverteilung-in-deutschland (1/1/2016).

Backhaus, W., S. Rupprecht and D. Franco 2019. *Road vehicle automation in sustainable urban mobility planning. Practitioner Briefing.* www.h2020-coexist.eu/wp-content/uploads/2019/06/ SUMP2.0_Practitioner-Briefings_Automation_Final-Draft.pdf (9/7/2019).

Bagloee, S. A., M. Tavana, M. Asadi and T. Oliver 2016. "Autonomous vehicles: Challenges, opportunities and future implications for transportation policies", in *Journal of Modern Transportation* (24) 4, 284–303.

Bagnasco, A., and P. Le Galès (eds.) 2000. *Cities in Contemporary Europe.* Cambridge: Cambridge University Press.

Bakici, T., E. Almirall and J. A. Wareham 2013. "Smart City Initiative: The Case of Barcelona", in *Journal of Knowledge Economies* (4) 2, 135–148. DOI:10.1007/s13132-012-0084-9.

Balducci, A., L. Boelens, J. Hillier, T. Nyseth and C. Wilkinson 2011. "Introduction: Strategic spatial planning in uncertainty: theory and exploratory practice", in *Town Planning Review* (82) 5, 481–501.

Bangemann, C. 2017. *Simulation einer urbanen Mobilitätslösung basierend auf autonom fahrenden E-Robotaxen in München.* Munich: Berylls Strategy Advisors.

Barnes, P., and E. Turkel 2017. *Autonomous Vehicles in Delaware: Analyzing the Impact and Readiness for the First State. Institute for Public Administration.* Newark, DE: University of Delaware.

BASt (Federal Highway Research Institute) 2012. *Rechtsfolgen zunehmender Fahrzeugautomatisierung.* www.bast.de/DE/Publikationen/Foko/Downloads/2012-11.pdf?__blob=publicationFile&v=1 (18/7/2019).

Bauer, H., P. Biwald, K. Mitterer and E. Thöni (eds.) 2017. *Finanzausgleich 2017: Ein Handbuch – mit Kommentar zum FAG 2017.* Öffentliches Management und Finanzwirtschaft 19. Vienna/Graz: NWV Neuer Wissenschaftlicher Verlag,

Bauernhansl, T., M. Ten Hompel and B. Vogel-Heuser (eds.) 2014. *Industrie 4.0 in Produktion, Automatisierung und Logistik: Anwendung-Technologien-Migration.* Wiesbaden: Springer Vieweg,

Bauriedl, S., and A. Strüver 2018. *Smart City – Kritische Perspektiven auf die Digitalisierung in Städten.* Bielefeld: transcript.

BBSR (Federal Institute for Research on Building, Urban Affairs and Spatial Development) 2010. "Leipzig Charta zur nachhaltigen europäischen Stadt", in *Informationen zur Raumentwicklung* 4/2010, 315–319.

Beck, U. 1997. *Was ist Globalisierung? Irrtümer des Globalismus – Antworten auf Globalisierung.* Frankfurt am Main: Suhrkamp.

Beck, U. 2013. *Risk Society: Towards a New Modernity.* Trans. Mark Ritter. Los Angeles, CA/London/New Delhi/Singapore/Washington D.C.: Sage.

Beck, U., W. Bonss and C. Lau 2003. "The Theory of Reflexive Modernization: Problematic, Hypotheses and Research Programme", in *Theory, Culture & Society* (20) 2, 1–33.

Beckmann, K. J. 2020. "Automatisierter Verkehr und Einsatz autonomer Fahrzeuge – (mögliche) Folgen für die Raumentwicklung", in Holz-Rau et al. (eds.) in *Wechselwirkungen von Mobilität und Raumentwicklung im Kontext gesellschaftlichen Wandels*, 244-269. Hannover.

Beecroft, R., H. Trenks, R. Rhodius, C. Benighaus and O. Parodi 2018. "Reallabore als Rahmen transformativer und transdisziplinärer Forschung: Ziele und Designprinzipien", in Di Giulio/Defila (eds.), 75–100.

Beiker, S. 2016. "Deployment Scenarios for Vehicles with Higher-Order Automation", in Maurer et al. (eds.), 197–211.

Beiker, S. 2018. "Deployment of automated driving as an example for the San Francisco Bay area", in Meyer/Beiker (eds.), 117–129.

Beirao, G., and J. A. Sarsfield-Cabral 2007. "Understanding attitudes towards public transport and private car: a qualitative study", in *Transport Policy* 14, 478–489.

Békési, S. 2005. "Verkehr in Wien. Personenverkehr, Mobilität und städtische Umwelt 1850 bis 2000", in Brunner/Schneider (eds.), 92–105.

Bemmann, M., B. Metzger and R. von Detten (eds.) 2014. *Ökologische Modernisierung. Zur Geschichte und Gegenwart eines Konzepts in Umweltpolitik und Sozialwissenschaften.* Frankfurt am Main/New York, NY: Campus.

Bento, N., and C. Wilson 2016. "Measuring the duration of formative phases for energy technologies", in *Environmental Innovation and Societal Transitions* 21, 95–112.

Berg, V. van den, and E. Verhoef 2016. "Autonomous cars and dynamic bottleneck congestion: The effects on capacity, value of time and preference heterogeneity", in *Transportation Research Part B: Methodological* 94, 43–60.

Berger, P. A., C. Keller, A. Klärner and R. Neef (eds.) 2014. *Urbane Ungleichheiten.* Wiesbaden: Springer VS.

Berking, H., and M. Löw (eds.) 2005. *Die Wirklichkeit der Städte, Soziale Welt, special vol. 16.* Baden-Baden: Nomos.

Berscheid, A. L. 2014. "Autonome Fahrzeuge und hegemoniale Männlichkeit in der Automobilkultur. Digitalisierung zwischen Utopie und Kontrolle", in *Femina Politica* 2/2014, 22–34.

Bertolini, L. 2007. "Evolutionary Urban Transportation Planning: An Exploration", in *Environment and Planning A: Economy and Space* (39) 8, 1998–2019.

Bertolini, L. 2012. "Integrating Mobility and Urban Development Agendas: A Manifesto", in *disP – The Planning Review* (48) 1, 16–26.

Bertolini, L. 2017. *Planning the Mobile Metropolis: Transport for People, Places and the Planet.* London: Palgrave/Red Grove Press.

Bertolini, L., C. Curtis and J. Renne 2012. "Station area projects in Europe and beyond: Towards transit oriented development?", in *Built Environment* (38) 1, 31–50.

Bloom, P., and A. Sancino 2019. *Disruptive Democracy: The Clash Between Techno-Populism and Techno-Democracy.* Thousand Oaks, CA: Sage.

BMF (Bundesministerium für Finanzen, Besteuerungsrechte und Abgabenerträge) 2018. *Besteuerungsrechte und Abgabenerträge.* Vienna: BMF. www.bmf.gv.at/budget/finanzbeziehungen-zu-laendern-und-gemeinden/besteuerungsrechte-und-abgabenertraege.html (14/8/2019).

BMUB (Federal Ministry for the Environment, Nature Conservation, Building and Nuclear Safety) 2015. *Umweltbewusstsein in Deutschland 2014. Ergebnisse einer repräsentativen Bevölkerungsumfrage.* Dessau-Roßlau: Umweltbundesamt (UBA). www.umweltbundesamt.de/sites/default/files/medien/378/publikationen/umweltbewusstsein_in_deutschland_2014.pdf (6/3/2019).

BMUB 2016. *Climate Action Plan 2050: Principles and goals of the German government's climate policy.* Berlin: BMUB. www.bmu.de/fileadmin/Daten_BMU/Pools/Broschueren/klimaschutzplan_2050_en_bf.pdf (24/11/2020) .

BMVBS (Bundesministerium für Verkehr, Bau und Stadtentwicklung) 2007. *Leipzig Charta zur nachhaltigen europäischen Stadt.* Berlin: BMVBS.

BMVI (Federal Ministry of Transport and Digital Infrastructure) 2018. *Automated and Connected Driving.* Berlin: BMVI. www.bmvi.de/EN/Topics/Digital-Matters/Automated-Connected-Driving/automated-and-connected-driving.html (19/3/2021).

BMVIT (Bundesministerium für Verkehr, Innovation und Technologie) 2016a. *Österreich unterwegs 2013/2014.* Vienna: BMVIT. www.bmvit.gv.at/verkehr/gesamtverkehr/statistik/oesterreich_unterwegs/downloads/oeu_2013-2014_Ergebnisbericht.pdf (28/8/2019).

BMVIT 2016b. *Automatisiert – Vernetzt – Mobil. Aktionsplan Automatisiertes Fahren*, June 2016. Vienna: BMVIT. www.bmvit.gv.at/service/publikationen/verkehr/automatisiert/downloadsautomatisiert2016.pdf (14/8/2019).

BMVIT 2016c. *Ergebnisbericht Projekt "ShareWay – Wege zur Weiterentwicklung von Shared Mobility zur dritten Generation".* Vienna: BMVIT.

BMVIT 2018. *Statistik Straße & Verkehr.* Vienna: BMVIT. www.bmvit.gv.at/verkehr/strasse/autostrasse/statistik/index.html (6/7/2019).

BMVIT 2019. *Was heißt Multimodalität? Österreich unterwegs, Teil 7.* BMVIT-Infothek. Vienna: BMVIT. https://infothek.bmvit.gv.at/teil-7-oesterreich-unterwegs/ (6/3/2019).

BMVIT (n.d.). *Elektromobilität.* Vienna: BMVIT. www.bmvit.gv.at/verkehr/elektromobilitaet/index.html (6/3/2019).

Boban, M., A. Kousaridas, K. Manolakis, J. Eichinger and W. Xu 2017. *Use Cases, Requirements, and Design Considerations for 5G V2X.* Ithaca, NY. https://arxiv.org/pdf/1712.01754.pdf (6/9/2019).

Bobeth, S., and E. Matthies 2016. "Elektroautos: Top in Norwegen, Flop in Deutschland? Empfehlungen aus Sicht der Umweltpsychologie", in *GAIA – Ecological Perspectives for Science and Society* (25) 1, 38–48. DOI:10.14512/gaia.25.1.10

Boeglin, J. 2015. "The Costs of Self-Driving Cars: Reconciling Freedom and Privacy with Tort Liability in Autonomous Vehicle Regulation", in *Yale Journal of Law & Technology* (17) 1, 171–203.

Böhler, S., D. Bongardt, C. Schäfer-Sparenberg and G. Wilke 2007. *Zukunft des Car-Sharing in Deutschland.* Wuppertal: Wuppertal Institut für Klima, Umwelt, Energie GmbH.

Bönninger, J., A. Eichelmann and U. Schüppel 2018. "Automatisiertes und vernetztes Fahren: Potenziale und Herausforderungen der technischen Entwicklung im Fahrzeug", in *Zeitschrift für Verkehrssicherheit* (64) 3, 97–98.

Boersma, R., B. van Arem and F. Rieck 2018. "Application of driverless electric automated shuttles for public transport in villages: the case of Appelscha", in *World Electric vehicle journal* (9) 1, 15–21.

Bösch, P. M. 2016. *Kapazitätsauswirkungen vollautonomer Fahrzeuge in der Schweiz.* 26th PTV Traffic Anwender Seminar, October 2016. Karlsruhe.

Bösch, P. M., F. Becker, H. Becker and K. W. Axhausen 2018. "Cost-based analysis of autonomous mobility services", in *Transport Policy* 64, 76–91.

Boltanski, L., and È. Chiapello 2001. "Die Rolle der Kritik in der Dynamik des Kapitalismus und der normative Wandel", in *Berliner Journal für Soziologie* (11) 4, 459–477.

Bormann, R., P. Fink, H. Holzapfel, S. Rammler, T. Sauter-Servaes, H. Tiemann, T. Waschke and B. Weirauch 2018. "Die Zukunft der Deutschen Automobilindustrie. Transformation by Disaster oder by Design?", in *WISO Diskurs 03/2018.* Bonn: Friedrich-Ebert-Stiftung.

Botsman, R. 2013. *The Sharing Economy Lacks A Shared Definition.* www.fastcompany.com/3022028/the-sharing-economy-lacks-a-shared-definition (6/3/2019).

Bourdieu, P. 1984. *Distinction: Social Critique of the Judgement of Taste.* Trans. R. Nice. London: Routledge Kegan & Paul.

Brake, K. 2011. "'Reurbanisierung': Globalisierung und neuartige Inwertsetzung städtischer Strukturen 'europäischen' Typs", in Frey/Koch (eds.), 299–323.

Brake, K., J. S. Dangschat and G. Herfert (eds.) 2001. *Suburbanisierung in Deutschland. Aktuelle Tendenzen.* Opladen: Leske + Budrich.

Brand, U., A. Demirovic, C. Görg and J. Hirsch (eds.) 2001. *Nichtregierungsorganisationen in der Transformation des Staates.* Münster: Westfälisches Dampfboot.

Braun-Thürmann, H. 2005. *Innovation.* Bielefeld: transcript. Brenner, M. 1985. "London", in Friedrichs (eds.), 149–254.

Brenner, N. 2004. "Urban governance and the production of new state spaces in western Europe, 1960–2000", in *Review of International Political Economy* (11) 3, 447–488.

Brenner, N., and R. Keil (eds.) 2006. *The global cities reader*. Abingdon: Routledge

Bröchler, S., and H.-J. Lauth (eds.) 2014. "Von Government zu Governance. Informales Regieren im Vergleich", in *Zeitschrift für Vergleichende Politikwissenschaft 8*, supplement 1, special issue 4. Wiesbaden: VS Verlag für Sozialwissenschaften.

Bröthaler, J., A. Haindl and K. Mitterer 2017. "Funktionsweisen und finanzielle Entwicklungen im Finanzausgleichssystem", in Bauer et al. (eds), 79–116.

Brundtland, G. H. 1987. *Our Common Future, Chapter 2: Towards Sustainable Development*. New York: UN.

Brunner, K., and P. Schneider (eds.). *Umwelt Stadt. Geschichte des Natur- und Lebensraumes Wien*. Vienna/Cologne/Weimar: Böhlau.

Bruns, F., M. Rothenfluh, M. Neuenschwander, M. Sutter, B. Belart and M. Egger 2018. *Einsatz automatisierter Fahrzeuge im Alltag: Denkbare Anwendungen und Effekte in der Schweiz*. Final report on Module 3c: "Mögliche Angebotsformen im kollektiven Verkehr (ÖV und ÖIV)". Basel: Basler Fonds.

Buchert, M. (ed.) 2014. *Reflexives Entwerfen – Reflexives Design*. Berlin: JOVIS.

Bukold, S. 2015. *Ölpreiskollaps, Verkehr & Klima. Daten und Strategien für den Klimagipfel in Paris*. Second part of the short study commissioned by the Bundestagsfraktion Bündnis 90/Die Grünen. www.gruene-bundestag.de/fileadmin/media/gruenebundestag_de/themen_az/klimaschutz/Bukold-OEl-Verkehr-Paris-30nov.pdf (6/3/2019).

BuroHappold Engineering 2016. *Urban Streets in the Age of connected and autonomous vehicles*. Global Design Sprint 2016. Bath. www.burohappold.com/wp-content/uploads/2016/11/BHE-Global-Design-Sprint16-v3.pdf (6/3/2019).

Busch-Geertsema, A., M. Lanzendorf, H. Müggenburg and H. Wilde 2016. "Mobilitätsforschung aus nachfrageorientierter Perspektive: Theorien, Erkenntnisse und Dynamiken des Verkehrshandelns", in Schwedes et al. (eds.), 755–779.

Campbell, M., M. Egerstedt, J. P. How and R. M. Murray 2010. "Autonomous driving in urban environments: approaches, lessons and challenges", in *Philosophical Transactions of the Royal Society A* (368), 4649–4672.

Cantos, P., M. Gumbau-Albert and J. Maudos 2005. "Transport infrastructures, spillover effects and growth: Evidence of the Spanish case", in *Transport Reviews* (25) 1, 25–50.

Canzler, W. 2010. "Mobilitätskonzepte der Zukunft und Elektromobilität", in Hüttl et al. (eds.), 39–61.

Canzler, W. 2015. "Zukunft der Mobilität: An der Dekarbonisierung kommt niemand vorbei", in *Aus Politik und Zeitgeschichte 41*. Frankfurt am Main: Societäts-Verlag. www.bpb.de/apuz/209960/zukunft-der-mobilitaet-an-der-dekarbonisierung-kommt-niemand-vorbei?p=all (20/11/2017).

Canzler, W., and A. Knie 2016. "Mobility in the age of digital modernity: Why the private car is losing its significance, intermodal transport is winning and why digitalisation is the key", in *Applied Mobilities* (1) 1, 56–67. DOI: 10.1080/23800127.2016.1147781.

Car Trottle 2017. *Testing The World's Smartest Autonomous Car (NOT A Tesla)*. Video. https://youtu.be/l3ELVACR2VY (28/2/2018).

Castells-Quintana, D., R. Ramos and V. Royuela 2015. "Income inequality in European Regions: Recent trends and determinants", in *Review of Regional Research (35)* 2, 123–146. DOI: 10.1007/s1003-7-015-0098-4.

Chan, N. D., and S. A. Shaheen 2012. "Ridesharing in North America: Past, Present, and Future", in *Transport Reviews (32)* 1, 93–112.

Chen, T. D., K. M. Kockelman and J. P. Hanna 2016. "Operations of a shared, autonomous, electric vehicle fleet: Implications of vehicle & charging infrastructure decisions", in *Transportation Research Part A: Policy and Practice 94*, 243–254.

Childress, S., B. Nichols, B. Charlton and S. Coe 2015. "Using An Activity-Based Model To Explore the Potential Impacts Of Automated Vehicles", in *Transportation Research Record (2493)* 1, 99–106.

Christensen, C. M. 2003. *The Innovator's Dilemma: When New Technologies Cause Great Firms to Fail*. New York, NY: HarperCollins.

Chui, M., M. Löffler and R. Roberts 2010. "The Internet of Things", in *The McKinsey Quarterly* (47) 2, 1–9. www.darley.com/documents/inside_darley/The_Internet_Of_Things_McKinsey_Report.pdf (12/1/2019).

City of San Francisco 2016. *Advanced Transportation and Congestion Management Technologies Deployment Initiative by SFMTA*. www.sfmta.com/sites/default/files/projects/2017/ ATCMTD%20 Grant%20Application.pdf (11/11/2018).

City of Vienna (Stadt Wien) 2014. *Smart City Wien: Framework Strategy*. www.wien.gv.at/stadtentwicklung/studien/pdf/b008384a.pdf (26/11/2020).

City Planning Authority 2014. *Development Strategy Göteborg 2035*. https://international.goteborg.se/sites/international.goteborg.se/files/ field_category_attachments/development_strategy_goteborg_2035.pdf (19/11/2018).

Clark, B. Y., N. Larco and R. F. Mann 2017. "The Impacts of Autonomous Vehicles and E-Commerce on Local Government Budgeting and Finance", in *SSRN Electronic Journal*. DOI:10.2139/ssrn.3009840.

Clarke, J., D. Bainton, N. Lendvai and P. Stubbs 2015. *Making policy move: Towards a politics of translation and assemblage*. Bristol: Policy Press.

CLC (Centre for Liveable Cities) 2014. *Liveable and Sustainable Cities – A Framework. Centre for Liveable and Sustainable Cities and Civil Service College*, Ministry of National Development (MND), Singapore. www.clc.gov.sg/docs/default-source/books/clc-csc-livea-ble-sustainable-cities.pdf (13/9/2018).

Clements, L. M., and K. M. Kockelman 2017. "Economic effects of automated vehicles", in *Transportation Research Record (2606)* 1, 106–114.

Coghlan, D., and M. Brydon-Miller (eds.) 2014, *The Sage Encyclopedia of Action Research*, Thousand Oaks, CA: Sage.

Crittenden, C. 2017. "A Drama in Time: How Data and Digital Tools are Transforming Cities and their Communities", in *City & Community* (16) 1, 3–8.

Curtin, R., Y. Shrago and J. Mikkelsen 2009. *Plug-in Hybrid Electric Vehicles*. University of Michigan Transportation Research Institute. http://evworld.com/library/UMich_PHEV_2009.pdf (6/3/2019).

Curtis, C., and J. Scheurer 2016. *Planning for Public Transport Accessibility. An International Sourcebook*. New York, NY: Routledge.

Cyganski, R., E. Fraedrich and B. Lenz 2015. "Travel-time valuation for automated driving: A Use-Case-driven Study", in *Proceedings of the 94th Annual Meeting of the TRB 2015*, Washington, DC.

Daimler 2018. *Fahrerlos geparkt. Automated Valet Parking*. www.daimler.com/innovation/case/autonomous/fahrerlos-geparkt.html (20/11/2018).

Dangschat, J. S. 1988. "Gentrification: Der Wandel innenstadtnaher Nachbarschaften", in Friedrichs (ed.), 272–292. DOI: 10.1007/978-3-322-83617-5_14.

Dangschat, J. S. 1992. "Konzeption, Realität und Funktion 'neuer Standortpolitik' – am Beispiel des 'Unternehmens Hamburg'", in Heinelt/Mayer (eds.), 29–48.

Dangschat, J. S. 2010. "Reurbanisierung: eine Renaissance der (Innen-)Städte?", in Frech/Reschl (eds.), 190–201.

Dangschat, J. S. 2014. "Soziale Ungleichheit und der (städtische) Raum", in Berger et al. (eds.), 117–132.

Dangschat, J. S. 2015a. "Gesellschaftliche Vielfalt: Heraus- oder Überforderung der Raumplanung?", in Dangschat et al. (eds.), 13–36.

Dangschat, J. S. 2015b. "Die geteilte Welt der Kommunikation. Wie das Web 2.0 die Stadt(teil)entwicklung verändert", in *Forum Wohnen und Stadtentwicklung 5/2015*, 245–250.

Dangschat, J. S. 2017a. "Automatisierter Verkehr: was kommt da auf uns zu?", in *Zeitschrift für Politische Wissenschaften 27*, 493–507. DOI:10.1007/s41358-017-0118-8.

Dangschat, J. S. 2017b. "Social Capital: Material for Social Bridging?", in Kapferer et al. (eds.), 40–60.

Dangschat, J. S. 2018. "Automatisierung und Vernetzung des (urbanen) Verkehrs: Neu-Erfindung oder Widerspruch zur 'Europäischen Stadt'?", in Gestring et al. (eds.), 313–335.

Dangschat, J. S. 2020. "Gesellschaftlicher Wandel, Raumbezug und Mobilität", in Holz-Rau et al. (eds.), in *Wechselwirkungen von Mobilität und Raumentwicklung im Kontext gesellschaftlichen Wandels* 244-269. Hannover.

Dangschat, J. S., M. Getzner, M. Haslinger and S. Zech (eds.) 2015. *Energie.Raum.Planung. Jahrbuch Raumplanung 2015.* Vienna: Neuer Wissenschaftlicher Verlag.

Dangschat, J. S., and A. Hamedinger 2009. "Planning Culture in Austria: The Case of Vienna, the Unlike City", in Knieling/Othengrafen (eds.), 95–112.

Datson, J. 2016. *Mobility as a Service: Exploring the Opportunity for Mobility as a Service in the UK.* Milton Keynes: Catapult Transport Systems. https://ts.catapult.org.uk/intelligent-mobility/im-resources/maasreport/ (6/3/2019).

Davis, S. C., S. W. Diegel and R. G. Boundy 2015. *Transportation energy data book* (no. ORNL-6984). Oak Ridge National Laboratory.

DCLG (Department for Communities and Local Government) 2015. *Ebbsfleet Garden City. Oral statement to parliament.* www.gov.uk/government/speeches/ebbsfleet-garden-city (24/2/2019).

DCLG (2016). *Locally-Led Garden Villages, Towns and Cities.* www.gov.uk/government/uploads/system/uploads/attachment_data/file/508205/Locally-led_garden_villages towns_and_cities.pdf (24/2/2019).

DCLG (2017). *First ever garden villages named with government support.* Press release from 2/1/2017. www.gov.uk/government/news/first-ever-garden-villages-named-with-government-support (24/2/2019).

Defila, R., and A. Di Giulio 2018. "Reallabore als Quelle für die Methodik transdisziplinären und transformativen Forschens: eine Einführung", in Di Giulio/Defila (eds.), 9–35.

Deloitte Development 2017a. *Autonomes Fahren in Deutschland: wie Kunden überzeugt werden.* Munich. www2.deloitte.com/content/dam/ Deloitte/de/Documents/consumer-industrial-products/Autonomes-Fahren-komplett-safe-Sep2016.pdf (25/2/2019).

Deloitte Development 2017b. *What's ahead for fully autonomous driving: Consumer opinions on advanced vehicle technology.* Munich. www2.deloitte.com/content/dam/Deloitte/us/Documents/manufacturing/us-manufacturing-consumer-opinions-on-advanced-vehicle-technology.pdf (24/4/2018).

Deltametropol 2013. *Maak Plaats! Wereken aan Knooppunkontwikkeling in Noord-Holland.* https://db.tt/sh9Oc48K (24/2/2019).

Deng, T. 2013. "Impacts of transport infrastructure on productivity and economic growth: Recent advances and research challenges", in *Transport Reviews (33)* 6, 686–699.

DETECTON Consulting 2016. Autonomes Fahren: Wenn das Lenkrad zur Sonderausstattung wird. Eine empirische Untersuchung der Akzeptanz autonom fahrender Fahrzeuge. Cologne: DETECTON.

DfT (Department for Transport) 2015. The Pathway to Driverless Cars: A detailed review of regulations for automated vehicle technology. London: Department for Transport. https://assets.publishing.service.gov.uk/government/uploads/system/uploads/attachment_data/file/401565/pathway-driverless-cars-main.pdf (21/11/2018).

Diehl, N., and C. Diehl 2018. "Autonomes Fahren im Diskurs: Semantische Netzwerke und diskursive Regelmäßigkeiten", in Siems/Papen (eds.), 325–340.

Diez, J. R., and S. Scholvin 2017. "Weltstädte des Globalen Südens in weltwirtschaftlichen Prozessen", in *Zeitschrift für Wirtschaftsgeographie (61)* 2, 61–64. DOI:10.1515/zfw-2017-0025.

Di Giulio, A., and R. Defila (eds.) 2018. *Transdisziplinär und transformativ forschen. Eine Methodensammlung.* Wiesbaden: Springer VS.

Dohnanyi, K. von 1983. "Unternehmen Hamburg". Speech in front of the Übersee-Club Hamburg on 29/11/1983. www.ueberseeclub.de/resources/Server/pdf-Dateien/1980-1984/vortrag-1983-11-29Dr.%20Klaus%20von%20Dohnanyi.pdf (29/10/2019).

Drive Sweden 2018. Website of Drive Sweden. www.drivesweden.net/en (19/11/2018).

DStT (Deutscher Städtetag) (ed.) 1971. *Rettet unsere Städte jetzt! Vorträge, Aussprachen und Ergebnisse der 16. Hauptversammlung des Deutschen Städtetages vom 25. bis 27. Mai 1971 in München.* Bonn: DStT.

Dueker, K.J., B. O. Bair and I. P. Levin 1977. "Ride-sharing: psychological factors", in T*ransportation Engineering Journal of the American Society of Civil Engineers* (103) 6, 685–692.

Dütschke, E., U. Schneider and A. Peters 2013. "Who will use electric vehicles?", in *Working Paper Sustainability and Innovation 6/2013.* Karlsruhe: Fraunhofer-Institut für System- und Innovationsforschung ISI.

Durand, A., L. Harms, S. Hoogendoorn-Lanser and T. Zijlstra 2018. *Mobility-as-a-Service and changes in travel preferences and travel behaviour: a literature review.* The Hague: KiM Netherlands Institute for Transport Policy Analysis.

Dusseldorp, M. 2017. *Zielkonflikte der Nachhaltigkeit: Zur Methodologie wissenschaftlicher Nachhaltigkeitsbewertungen.* Wiesbaden: J. B. Metzler.

e:mobil 2018. *e-Carsharing: Das gibt es in den Bundesländern.* Vienna. https://elektrotechnikblog.at/e-carsharing-das-gibt-es-in-den- bundeslaendern/ (4/11/2019).

Ebert, G., P. Behrens, M. Landau, T. Pregger and M. Specht 2012. *Integration von Elektromobilen in das Smart Grid: Intelligente Beladung von Elektrofahrzeugen, FVEE – Themen 2012.* Berlin. www.fvee.de/fileadmin/publikationen/Themenhefte/th2012-2/th2012_08_04.pdf (6/3/2019).

Eckhardt, J., A. Aapaoja, L. Nykänen, J. Sochor, M. A. Karlsson and D. König 2018. "The European Roadmap 2025 for Mobility as a Service", in *Proceedings of 7th Transport Research Arena TRA 2018*, Vienna.

Eckstein, L., T. Form, M. Maurer, R. Schöneburg, G. Spiegelberg and C. Stiller 2018. *Automatisiertes Fahren.* VDI status report, July 2018. Dusseldorf.

Egbue, O., and S. Long 2012. "Barriers to widespread adoption of electric vehicles: An analysis of consumer attitudes and perceptions", in *Energy Policy* 48, 717–729.

Eimler, S. C., and S. Geisler 2015. "Zur Akzeptanz des Autonomen Fahrens: eine A-Priori Studie", in Weisbecker et al. (eds.), 533–540.

Elbanhawi, M., M. Simic and R. Jazar 2015. "Improved manoeuvering of autonomous passenger vehicles: Simulations and field results", in *Journal of Vibration and Control* (21) 13, 1–30.

Elliot, C. S., and G. Long 2016. "Manufacturing rate busters: computer control and social relations in the labour process", in *Work, Employment and Society* (30) 1, 135–151.

Engels, F. 1887. *The Condition of the Working Class in England in 1844.* New York, NY: J.W. Lovell Co..

EPOMM – European Platform on Mobility Management 2017. *The Role of Mobility as a Service in Mobility Management, e-update, December 2017.* http://epomm.eu/newsletter/v2/content/2017/1217_2/doc/eupdate_en.pdf (20/8/2020).

European Commission 2005. *Trans-European Transport Network: TEN-T priority axes and projects 2005.* Luxemburg: Office for Official Publications of the European Communities. https://ec.europa.eu/ten/transport/projects/doc/2005_ten_t_en.pdf

European Commission 2016. *Communication from the commission to the European parliament, the council, the European economic and social committee and the committee of the regions.* https://ec.europa.eu/transport/sites/transport/files/com20160766_en.pdf (3/9/2018).

European Commission 2018. *Europe on the Move. Sustainable Mobility for Europe: safe, connected, and clean.* Communication from the Commission to the European Parliament, the Council, the European Economic and Social Committee and the Committee of the Regions. https://eur-lex.europa.eu/legal-content/EN/TXT/HTML/?uri=CELEX:52018DC0293&from=DE (20/8/2020).

European Commission 2017. *C-ITS Platform: Final report.* Brussels. https://ec.europa.eu/transport/sites/transport/files/2017-09-c-its-platform-final-report.pdf (25/2/2019).

European Commission 2018. *STRIA (Strategic Transport Research and Innovation Agenda) 2.0 Roadmap. Connected and Automated Transport – Road – Draft Version 11.1, 11/11/2018.* Brussels.

European Commission 2019. *STRIA Roadmap on Connected and Automated Transport: Road, Rail and Waterborne.* Brussels. http://ec.europa.eu/research/transport/pdf/stria/stria-roadmap_on_connected_and_automated_transport2019-TRIMIS_website.pdf (10/4/2019).

European Environment Agency 2017. *Annual European Union greenhouse gas inventory 1990–2015 and inventory report 2017. Submission to the UNFCCC Secretariat, May 2017.* Copenhagen. www.eea.europa.eu/publications/european-union-green-house-gas-inventory2017/download (22/7/2019).

European Union 2011. *Cities of Tomorrow: Challenges, Visions, Ways Forward.* Brussels: European Commission, DG for Regional Policy. https://ec.europa.eu/regional_policy/sources/docgener/studies/pdf/citiesoftomorrow/citiesoftomorrow_final.pdf (19/3/2021).

Eurostat 2017. *Population and population change, capital cities.* Brussels. http://ec.europa.eu/eurostat/statistics-explained/index.php?title=File:Population_and_population_change_capital_cities_2004%E2%80%932014_RYB17.png#file (24/2/2019).

Ernst & Young 2013. *Autonomes Fahren: die Zukunft des Pkw-Marktes? Was Autofahrer von Fahrzeugen mit Autopilot halten und wie sie über Fahrerassistenzsysteme denken.* Eschborn: Ernst & Young.

Ernste, H., and V. Meier (eds.) *1992. Regional Development and Contemporary Response: Extending Flexible Specialization.* London: Belhaven Press.

Färber, B. 2016. "Communication and Communication Problems Between Autonomous Vehicles and Human Drivers", in: Maurer et al. (eds.), 125–144.

Fagnant, D. J., and K. M. Kockelman 2014. "The travel and environmental implications of shared autonomous vehicles, using agent-based model scenarios", in *Transportation Research Part C: Emerging Technologies 40*, 1–13.

Fagnant, D. J., and K. M. Kockelman 2015. "Preparing a Nation for Autonomous Vehicles: Opportunities, Barriers and Policy Recommendations for Capitalizing on Self-Driven Vehicles", in *Transport Research Part A: Policy and Practice 77*, 167–181. DOI:10.1016/j.tra.2015.04.003.

Fagnant, D. J., K. M. Kockelman and P. Bansal 2015. "Operations of shared autonomous vehicle fleet for Austin, Texas, market", in *Transportation Research Record (2563)* 1, 98–106.

Favarò, F. M., S. Eurich and N. Nader 2018. "Autonomous vehicles' disengagements: Trends, triggers, and regulatory limitations", in *Accident Analysis and Prevention 110*, 136–148.

Fellendorf, M. 2018. *Straßenbauliche Infrastruktur und VLSA für automatisierte Fahrzeuge. Presentation in the context of the workshop on the research project AUTO-NOM.* Vienna.

Fend L., and J. Hofmann (eds.) *2018. Digitalisierung in Industrie-, Handels- und Dienstleistungsunternehmen. Konzepte – Lösungen – Beispiele.* Wiesbaden: Springer Gabler.

Fernald, J. G. 1999. "Roads to prosperity? Assessing the link between public capital and productivity", in *American Economic Review (89)* 3, 619–638.

FGSV – Road and Transport Research Association 2006 (trans. 2012). *Directives for the Design of Urban Roads,* RASt 06. Cologne.

FIRST 2018. *Funding Program for World-Leading Innovative R&D on Science and Technology (FIRST).* Presentation by the Bureau of Science, Technology and Innovation, Cabinet Office, Government of Japan. Tokyo. www.m-e-f.info/app/download/.../0426_0915_Mr.%20Shin- do_MEF_2017.pdf (22/6/2019).

Fischedick, G., and A. Grunwald (eds.) 2017. *Pfadabhängigkeiten in der Energiewende: Das Beispiel Mobilität, Energiesysteme der Zukunft series.* Munich: acatech.

Flämig, H. 2016. "Autonomous Vehicles and Autonomous Driving in Freight Transport", in Maurer et al. (eds.), 365–385.

Flander, K. de, U. Hahne, H. Kegler, D. Lang, R. Lucas, U. Schneidewind, K.-H. Simon, M. Singer-Brodowski, M. Wanner and A. Wiek 2014. "Resilienz und Reallabore als Schlüsselkonzepte urbaner Transformationsforschung. Zwölf Thesen", in *GAIA – Ecological Perspectives for Science and Society (23)* 3, 284–286.

Forest, A., and M. Konca 2007. *Autonomous Cars and Society. Worcester Polytechnic Institute.* https://web.wpi.edu/Pubs/E-project/Available/E-project-043007-205701/un-restricted/IQPOVP06B1.pdf (31/5/2017).

Fraade-Blanar, L., M. S. Blumenthal, J. M. Anderson and N. Kalra 2018. *Measuring Automated Vehicle Safety.* Santa Monica, CA: RAND Corporation.

Fraedrich, E., and B. Lenz 2016a. "Societal and Individual Acceptance of Autonomous Driving", in Maurer et al. (eds.), 621–640.

Fraedrich, E., and B. Lenz 2016b. "Taking a Drive, Hitching a Ride: Autonomous Driving and Car Usage", in Maurer et al. (eds.), 665–685.

Fraedrich, E., R. Cyganski, I. Wolf and B. Lenz 2016. *User Perspectives on Autonomous Driving. A Use-Case-Driven Study in Germany.* Humboldt-Universität zu Berlin, Geografisches Institut, progress report no. 187. Berlin: HUB.

Fraedrich, E., D. Heinrichs, F. J. Bahamonde-Birke and R. Cyganski 2018. "Autonomous driving, the built environment and policy implications", in *Transportation Research Part A: Policy and Practice 122.* DOI:10.1016/j.tra.2018.02.018

Fraser, N., and L. Gordon 1994. "Civil citizenship against social citizenship? On the ideology of contract-versus-charity", in Steenbergen (ed.), 90–107.

Fraunhofer ISI (Fraunhofer Institut für System- und Innovationsforschung) 2012. *Roadmap zur Kundenakzeptanz. Zentrale Ergebnisse der sozialwissenschaftlichen Begleitforschung in den Modellregionen. Technologie-Roadmapping am Fraunhofer ISI: Konzepte – Methoden – Praxisbeispiele Nr. 3.* www.isi.fraunhofer.de/content/dam/isi/dokumente/cce/2012/road-map_broschuere_netz.pdf (6/3/2019).

Frech, S., and R. Reschl (eds.) 2010. *Urbanität neu planen. Stadtplanung, Stadtumbau, Stadtentwicklung.* Bad Schwalbach: Wochenschau-Verlag.

Freese, C., and T. Schönberg 2014. *Shared Mobility: How new businesses are rewriting the rules of the private transportation game.* Munich: Roland Berger.

Freudendal-Pedersen, M., K. Hartmann-Petersen, A. Kjaerulff and L. D. Nielsen 2017. "Interactive environmental planning: creating utopias and storylines with a mobilities planning project", in *Journal of Environmental Planning and Management (60)* 6, 941–958.

Freudendal-Pedersen, M., and S. Kesselring 2016. "Mobilities, Futures & the City: repositioning discourses – changing perspectives – rethinking policies", in *Mobilities (11)* 4, 575–586.

Freudendahl-Pedersen, M., S. Kesselring and E. Servou 2019. "What is Smart for the Future City? Mobility and Automation", in *Sustainability (11)* 1, 221. DOI: 10.3390/su11010221.

Frey, O., and F. Koch (eds.) 2011. *Die Zukunft der Europäischen Stadt.* Wiesbaden: VS Verlag für Sozialwissenschaften.

Freytag, T., H. Gebhardt, U. Gerhard and D. Wastl-Walter (eds.) 2016. *Humangeographie kompakt.* Berlin/Heidelberg: Springer Spektrum.

Friedrich, B. 2016. "The Effect of Autonomous Vehicles on Traffic", in Maurer et al. (eds.), 317–334.

Friedrich, M., and M. Hartl 2016. *Modellergebnisse geteilter autonomer Fahrzeugflotten des öffentlichen Nahverkehrs. Final report MEGAFON.* Stuttgart: Universität Stuttgart, Institut für Straßen- und Verkehrswesen.

Friedrichs, J. 1978. "Steuerungsmaßnahmen und Theorie der Suburbanisierung", in *Akademie für Raumforschung Landesplanung* (ed.), 15–33.

Friedrichs, J. (ed.) 1985. *Stadtentwicklungen in West- und Osteuropa.* Berlin/New York, NY: Walter de Gruyter.

Friedrichs, J. (ed.) 1988. *Soziologische Stadtforschung. Kölner Zeitschrift für Soziologie und Sozialpsychologie, special issue no. 29.* Wiesbaden: VS Verlag für Sozialwissenschaften.

Fuller, B. 2016. "Cautious Optimism about Driverless Cars and Land Use in American Metropolitan Areas", in *Cityscape: A Journal of Policy Development and Research (18)* 3, 181–184.

Funabashi, Y. (ed.) 2018. *Japan's Population Implosion: The 50 million shock.* New York, NY: Palgrave Macmillan.

Geels, F. W. 2005. "The Dynamics of Transitions in Socio-technical Systems: A Multi-level Analysis of the Transition Pathway from Horse-drawn Carriages to Automobiles (1860–1930)", in *Technology Analysis & Strategic Management (17)* 4, 445–476.

Geels, F. W. 2011. "The multi-level perspective on sustainability transitions: Responses to seven criticisms", in *Environmental Innovation and Societal Transitions (1)* 1, 24–40. DOI: 10.1016/j.eist.2011.02.002.

Gertz, C., and M. Dörnemann 2016. *Wirkungen des autonomen/fahrerlosen Fahrens in der Stadt. Entwicklung von Szenarien und Ableitung der Wirkungsketten.* Bremen.

Gestring, N., and J. Wehrheim (eds.) 2018. *Urbanität im 21. Jahrhundert. Eine Fest- und Freundschaftsschrift für Walter Siebel.* Frankfurt am Main/New York, NY: Campus.

Getzner, M., and J. Kadi 2019. "Determinants of land consumption in Austria and the effects of spatial planning regulations", in *European Planning Studies*. DOI: 10.1080/09654313.2019.1604634.

Geurs, K. T., and B. van Wee 2004. "Accessibility evaluation of land use and transport strategies: review and research directions", in *Journal of Transport Geography 12*, 127–140.

Giersig, N. 2005. *Urban governance and the European City: Illustrating the interconnectedness of two contemporary debates.* RTN Urban Europe Working Paper, Universitá di Urbino. www.urban-europe.net/working/04_2005_Giersing.pdf (12/2/2019).

Giffinger, R., J. S. Dangschat and J. Suitner 2018. "Zur Notwendigkeit der raumbezogenen Forschung zu digitalen Transformationsprozessen", in Suitner et al. (eds.), 7–21.

GLA (Greater London Authority) 2015. *London Infrastructure 2050. Visual depiction by Greater London Authority & The Mayor of London.* www.london.gov.uk/sites/default/files/london_infrastructure_2050_e-book.pdf (21/11/2018).

GLA 2017. *The London Plan. The Spatial Development Strategy for Greater London.* Draft for public consultation. www.london.gov.uk/sites/default/files/new_london_plan_december_2017.pdf (21/11/2018).

Glaser, S., B. Vanholme, S. Mammar, D. Gruyer and L. Nouveliere 2010. "Maneuver-based trajectory planning for highly autonomous vehicles on real road with traffic and driver interaction", in *IEEE Transactions on Intelligent Transportation Systems* (11) 3, 589–606.

Godefrooij, T. 2012. *Integration of Cycling & Public Transport in the Netherlands.* Dutch Cycling Embassy. www.trm.dk/~/media/files/publication/2012/traeng-selskommission/konference-den-1-oktober-2012/3-tom-godefrooij.pdf (2/2/2018).

Göderitz, J., R. Rainer and H. Hoffmann 1957. *Die gegliederte und aufgelockerte Stadt*, Archiv für Städtebau und Landesplanung 4. Tübingen: Wasmuth.

Goldzamt, E. 1973. *Städtebau sozialistischer Länder. Soziale Probleme.* Berlin: VEB Verlag für Bauwesen.

Gontar, P., S. A. E. Schneider, C. Schmidt-Moll, C. Bollin and K. Bengler 2017. "Hate to interrupt you, but … analyzing turnarounds from a cockpit perspective", in *Cognition, Technology & Work* (19) 4, 837–853.

Gossen, M. 2012. *Nutzen statt Besitzen. Motive und Potenziale der internetgestützten gemeinsamen Nutzung am Beispiel des Peer-to-Peer Car-Sharing,* IÖW series 202/12. Berlin: Institut für ökologische Wirtschaftsforschung.

Goulding, R., and M. Kamargianni 2018. *The Mobility as a Service Maturity Index: Preparing the Cities for the Mobility as a Service Era.* Proceedings of the 7th Transport Research Arena TRA 2018, Vienna.

Graham, S. (ed.) 2004. *The Cybercities Reader.* London: Routledge.

Grant, R., and J. Nijman 2006. "Globalization and the Corporate Geography of Cities in the Less-Developed World", in Brenner/Keil (eds.), 224–237.

Grin, J. 2006. "Reflexive Modernisation as a Governance Issue, Or: Designing and Shaping Re-structuration", in Voß et al. (eds.), 54–81.

Groß, M., H. Hoffmann-Riem and W. Krohn 2005. *Realexperimente: Ökologische Gestaltungsprozesse in der Wissensgesellschaft.* Bielefeld: transcript.

Grubler, A., C. Wilson and G. Nemet 2016. "Apples, oranges, and consistent comparisons of the temporal dynamics of energy transitions", in *Energy Research & Social Science 22*, 18–25.

Grush, B., J. Niles and E. Baum 2016. *Ontario Must Prepare for Vehicle Automation Automated vehicles can influence urban form, congestion and infrastructure delivery.* Ontario: Residential and Civil Construction Alliance.

Guerra, E. 2016. "Planning for Cars That Drive Themselves: Metropolitan Planning Organizations, Regional Transportation Plans, and Autonomous Vehicles", in *Journal of Planning Education and Research* (36) 2, 210–224.

Guerra, E., and E. A. Morris 2018. "Cities, Automation, and the Self-parking Elephant in the Room", in *Planning Theory & Practice* (19) 2, 291–297.

Gugler, J. 2004. *World cities beyond the west. Globalization, development, and inequality.* Cambridge: Cambridge University Press.

Guthrie, K. K., and W. H. Dutton 1992. "The Politics of Citizen Access Technology", in *Policy Studies Journal* (20) 4, 574–597.

Haahtela, T., and E. Viitamo 2017. *Searching for the potential of MaaS in commuting – comparison of survey and focus group methods and results.* Paper presented at the 1st International Conference on Mobility-as-a-Service, Tampere.

Habermas. J. 1988. *Legitimation Crisis.* Trans. T. McCarthy. Cambridge: Polity.

Hämer, H.-W. 1990. "Behutsame Stadterneuerung", in *Senatsverwaltung für Bau- und Wohnungswesen* (ed.).

Häußermann, H. 2001. "Die Europäische Stadt" in Leviathan (29) 2, 237–255.

Häußermann, H. 2005. "The End of the European City?", in *European Review* (13) 2, 237–249. DOI: 10.1017/ S1062798705000372.

Häußermann, H., and A. Haila 2005. "The European City: A Conceptual Framework and Normative Project", in Kazepov (ed.), 43–64.

Haferburg, C., and J. Oßenbrügge 2009. "Die neue Corporate Geography in den Global Cities des Südens: Das Beispiel Johannesburg", in Altrock et al. (eds.), 29–45.

Haider, T., and R. Klementschitz 2017. *Wirkungspotentiale für den Einsatz automatisierter Fahrzeuge im ländlichen Raum.* Ergebnisbericht Forschungsprojekt "Shared Autonomy", Institut für partizipative Sozialforschung, Universität für Bodenkultur. Vienna.

Hajer, M. 2014. "On Being Smart about Cities: Seven Considerations for a New Urban Planning and Design", in Hajer/Dassen (eds.), 11–43.

Hajer, M. 2016. "On Being Smart abut Cities: Seven Considerations for a New Urban Planning and Design", in: Allen et al. (eds.), 50–63.

Hajer, M., and T. Dassen (eds.) 2014. *Smart About Cities: Visualising the Challenge for 21st Century Urbanism.* Rotterdam: PBL publishers.

Hall, P., and K. Pain 2006. *The polycentric metropolis. Learning from mega-city regions in Europe.* London: Routledge.

Halpern, O., J. LeCavalier, N. Calvillo and W. Pietsch 2013. "Test-Bed Urbanism", in *Public Culture* (25) 2, 272–306.

Hamedinger, A. 2013. *Governance, Raum und soziale Kohäsion: Aspekte einer sozial kohäsiven stadtregionalen Governance.* Habilitation thesis, Fakultät für Architektur und Raumplanung, TU Wien, Vienna.

Hammer, K. (ed.) 2016, "Wien wächst – Smart City. Neues Konzept, offene Fragen", in *Standpunkte Nr. 22.* Vienna: Arbeiterkammer Wien.

Hancock, P. A., D. R. Billings, K. E. Schaefer, J. Y. Chen, E. J. De Visser and R. Parasuraman 2011. "A metaanalysis of factors affecting trust in human-robot interaction", in *Human Factors* (53) 5, 517–527.

Hanson, G. 2015. *The UK Corridor – A2/M2 Connected Vehicle Corridor.* https://amsterdamgroup.mett.nl/Downloads/downloads_get-filem.aspx?id=506552 (24/2/2019).

Harms, S. 2003. *Besitzen oder Teilen. Sozialwissenschaftliche Analyse des Car-Sharings.* Zurich: Rüegger.

Harms, L., A. Durand, S. Hoogendoorn-Lanser, and T. Zijlstra 2018. *Exploring Mobility-as-a-Service. Insights from Literature and Focus Group Meetings.* The Hague: Netherlands Institute for Transport Policy Analysis (KiM).

Harvey, D. 1989. "From Managerialism to Entrepreneurialism: The Transformation in Urban Governance in Late Capitalism. Geografiska Annaler. Series B", in *Human Geography* (71) 1, 3–17. DOI:10.2307/ 490503.

Hassenpflug, D. (ed.) 2002. *Die Europäische Stadt – Mythos und Wirklichkeit.* Münster: LIT.

Hawkins, A. J. 2017. "Waymo is first to put fully self-driving cars on US roads without a safety driver". www.theverge.com/2017/11/7/16615290/waymo-self-driving-safety-driver-chandler-autonomous (24/2/2019).

Heeg, S., and M. Rosol 2007. "Neoliberale Stadtpolitik im globalen Kontext. Ein Überblick" in *PROKLA* (149) 4, 491–510.

Heidbrink, L. 2007. *Handeln in der Ungewissheit: Paradoxien der Verantwortung.* Berlin: Kulturverlag Kadmos.

Heinecke, A. 2012. *Generelle Anmerkungen zur Szenario-Technik.* Braunschweig: mimeo. www.sinus-online.com/images/Anmerkungen_zur_Szenario-Technik.pdf (2/2/2019).

Heinelt, H., and M. Mayer (eds.) 1992. *Politik in europäischen Städten. Fallstudien zur Bedeutung lokaler Politik.* Stadtforschung aktuell 38. Basel: Birkhäuser.

Heinrichs, D. 2016. "Autonomous Driving and Urban Land Use", in Maurer et al. (eds.), 213–231.

Heinrichs, D., S. Rupprecht and S. Smith 2019. "Making Automation Work for Cities: Impacts and Policy Responses", in Meyer/Beiker (eds.), 243–252.

Hellåker, J., J. Gunnarson and P. King 2019. "Drive Sweden: An Update on Swedish Automation Activities", in Meyer/Beiker (eds.), 41–49.

Hern, A. 2018. "First robot delivery drivers start work at Silicon Valley campus", *The Guardian*, 30/4/2018, www.theguardian.com/cities/2018/apr/30/robot-delivery-drivers-coming-to-a-campus-near-you-starship-technologies (28/8/2019).

Heyen, D. A., B. Brohmann, J. Libbe, R. Riechel and J. H. Trapp 2018. *Stand der Transformationsforschung unter besonderer Berücksichtigung der kommunalen Ebene. Paper in the context of the project* "Vom Stadtumbau zur städtischen Transformationsstrategie" in the research programme "Experimenteller Wohnungs- und Städtebau" (Ex-WoSt). Berlin: Öko-Institut & Deutsches Institut für Urbanistik (Difu).

Hidrue, M. K., G. R. Parsons, W. Kempton, and M. P. Gardner 2011. "Willingness to pay for electric vehicles and their attributes", in *Resource and Energy Economics* (33) 3, 686–705.

Hildebrandt, A., and W. Landhäußer (eds.) 2017. *CSR und Digitalisierung. Der digitale Wandel als Chance und Herausforderung für Wirtschaft und Gesellschaft.* Berlin: Springer.

Hillier, J. 2016. "Strategic spatial planning in uncertainty or planning indeterminate futures? A critical review", in Albrechts et al. (eds.), 298–316.

Ho, C., D. A. Hensher, C. Mulley and Y. Wong 2017. *Prospects for switching out of conventional transport services to mobility as a service subscription plans – A stated choice study.* Paper presented at the International Conference Series on Competition and Ownership in Land Passenger Transport (Thredbo 15). Stockholm.

Hoadley, S. (ed.) 2017. *Mobility as a service: Implications for urban and regional transport. Discussion paper offering the perspective of Polis member cities and regions on Mobility as a Service (MaaS).* Polis Traffic Efficiency & Mobility Working Group, September 2017. www.polisnetwork.eu/uploads/Modules/PublicDocuments/polis-maas-discussion-paper-2017---final_.pdf (6/3/2019).

Hoff, K. A., and M. Bashir 2015. "Trust in automation: Integrating empirical evidence on factors that influence trust", in *Human Factors* (27) 3, 407–434.

Hollestelle, M. L. 2018. *Automated Driving: Driving urban development? An integrated modelling and research-by-design approach on the spatial impacts of automated driving.* Master's thesis, TU Delft.

Holmberg, P.-E., M. Collado, S. Sarasini and M. Williander 2016. *Mobility as a Service – MaaS: Describing the framework.* Stockholm: Viktoria Swedish ICT AB.

Holz-Rau, C., U. Reutter and J. Scheiner (eds.) 2020. *Wechselwirkungen von Mobilität und Raumentwicklung im Kontext des gesellschaftlichen Wandels*, Akademie für Raumforschung und Landesplanung (ARL), Forschungsberichte der ARL X. Hannover: ARL.

Honneth, A. 2016. "Organisierte Selbstverwirklichung. Paradoxien der Individualisierung", in: Menke/Rebentisch (eds.), 63–80.

Hoogendoorn, R., B. van Arem and S. Hoogendoorn 2014. "Automated Driving, Traffic Flow Efficiency, and Human Factors: Literature Review", in *Transportation Research Record 2422*, 113–120.

Hopkins, D., and T. Schwanen 2018. "Automated Mobility Transitions: Governing Processes in the UK", in *Sustainability* (10) 4, 956.

Horizont & Bundesverband CarSharing. (2019). "Anzahl registrierter Carsharing-Nutzer in Deutschland in den Jahren 2008 bis 2019", in Statista. https://de.statista.com/statistik/daten/studie/324692/umfrage/carsharing-nutzer-in-deutschland/ (28/8/2019).

Hradil, S. 1995. *Die Single-Gesellschaft.* Munich: Beck.

Hülsmann, F., J. Wiepking, W. Zimmer, G. Sunderer, K. Götz and Y. Sprinke 2018. *share – Wissenschaftliche Begleitforschung zu car2go mit batterieelektrischen und konventionellen Fahrzeugen. Forschung zum free-floating Carsharing.* Berlin/Frankfurt am Main: Öko-Insitut e. V./ISOE – Institut für sozial-ökologische Forschung. www.isoe-publikationen.de/fileadmin/redaktion/Projekte/share/share_Endbericht.pdf (22/8/2019).

Hüttl, R. F., B. Pischetsrieder and D. Spath (eds.) 2010. *Elektromobilität – Potenziale und wissenschaftlich-technische Herausforderungen.* Potsdam: Helmholtz Zentrum.

Huiling E., and B. Goh 2017. "AI, Robotics and Mobility as a Service: The Case of Singapore", in *Field Actions Science Report 17*, 26–29. http://journals.openedition.org/factsreports/4411 (13/8/2018).

Human Resources (2018). *Singapore to potentially suffer talent shortage of more than 1m workers.* www.humanresourcesonline.net/singapore-to-potentially-suffer-talent-shortage-of-more-than-1m-workers/ (13/9/2019).

Husted, M., and D. M. B. Tofteng 2014. "Critical Utopian Action Research", in Coghlan/Brydon-Miller (eds.), 230–232.

Huxley, M. 2002. "Governmentality, Gender, Planning: A Foucauldian Perspective", in Allmendinger/Tewdwr-Jones (eds.), 136–153.

Ingenieur.de 2018. *Unüberhörbar: Elektroautos bekommen künstlichen Sound. E-Autos sind zu leise.* www.ingenieur.de/technik/forschung/unueberhoerbar-elektro-autos-bekommen-kuenstlichen-sound/ (6/3/2019).

Innes, J. E., and D. E. Booher 2010. *Planning with complexity: An introduction to collaborative rationality for public policy.* Abingdon/New York, NY: Routledge.

IPSS (National Institute of Population and Social Security Research) 2017. *Population Projections for Japan 2016 to 2065.* http://fpcj.jp/wp/wp-content/uploads/2017/04/1db9de3ea4ade06c3023d-3ba54dd980f.pdf (13/10/2018).

Jäger, S. 2015. *Kritische Diskursanalyse. Eine Einführung*, 7th ed. Münster: UNRAST.

Jahn, M., H. Maas, J, N, Ries, D. Wagner, S. Braun and M. Feldwieser 2017. *Digitalisierung und die Transformation des urbanen Akteursgefüges.* Bonn: Bundesinstitut für Bau-, Stadt- und Raumforschung im Bundesamt für Bauwesen und Raumordnung.

Jann, W., M. Röber and H. Wollmann 2006. *Public Management – Grundlagen, Wirkungen und Kritik.* Berlin: Edition Sigma.

Jessop, B. 1992. "Post-Fordism and flexible specialization: Incommensurable, contradictory, complementary, or just plain different perspectives?", in Ernste/Meier (eds.), 25–43.

Jessop, B. 2002. *The Future of the Capitalist State.* Cambridge: Polity.

Jessop, B. 2003. *Governance and Metagovernance: On Reflexivity, Requisite Variety, and Requisite Irony.* Lancaster: Lancaster University, Department of Sociology.

Jeute, G. H. 2017. "Zur Frage einer Globalisierung im Mittelalter im Hinblick auf transkontinentale Verflechtungen", in *Mitteilungen der Deutschen Gesellschaft für Archäologie des Mittelalters und der Neuzeit 30*, 25–32. DOI: 10.11588/dgamn.2017.0.40246.

Jittrapirom, P., V. Caiati, A. M. Feneri, S. Ebrahimigharehbaghi, M. J. Alonso-González and J. Narayan 2017. "Mobility as a service: A critical review of definitions, assessments of schemes, and key challenges", *Urban Planning (2) 2*, 13–25.

Jiwattanakulpaisarn, P., R. B. Noland and D. J. Graham 2012. "Marginal productivity of expanding highway capacity", in *Journal of Transport Economics and Policy* (46) 3, 333–347.

Johannig, V., and R. Mildner 2015. *Car IT kompakt. Das Auto der Zukunft – Vernetzt und autonom fahren.* Wiesbaden: Springer.

Jones, P. M. 2017. "The evolution of urban transport policy from car-based to people-based cities: Is this development path universally applicable?", in *Proceedings of the 14th World Conference on Transport Research.* Shanghai.

Joschunat, H., A. Knie and L. Ruhrort 2016. *Zukunftsfenster in eine disruptive Mobilität. 1. InnoZ-Zukunftsfenster: Disruptive Transformation der Mobilitätswelt. Teil 1: Mobilität in einer vernetzten Welt.* InnoZ – Innovationszentrum für Mobilität und gesellschaftlichen Wandel. Berlin: InnoZ.

JSPS (Japan Society for the Promotion of Science) 2018. *Funding Program for World-Leading Innovative R&D on Science and Technology.* Presentation of Bureau of Science, Technology and Innovation, Cabinet Office of Japan. www.jsps.go.jp/english/e-first/index.html (12/10/2018).

Kaelble, H. 2001. "Die Besonderheit der europäischen Stadt im 20. Jahrhundert", in *Leviathan* (29) 2, 256–274.

Kagermann, H. 2017. "Die Mobilitätswende: Die Zukunft der Mobilität ist elektrisch, vernetzt und automatisiert", in Hildebrandt/Landhäußer (eds.), 357–371.

Kagermeier, A. (1997). *Siedlungsstruktur und Verkehrsmobilität. Eine empirische Untersuchung am Beispiel von Südbayern.* Dortmund: Dortmunder Vertrieb für Bau- und Planungsliteratur.

Kainrath, W. 1997. *Die Bandstadt. Städtebauliche Vision oder reales Modell der Stadtentwicklung.* Vienna: Picus.

Kamargianni, M., M. Matyas and W. Li 2018. *Londoners' attitudes towards car-ownership and Mobility-as-a-Service: Impact assessment and opportunities that lie ahead.* MaaSLab – UCL Energy Institute Report. London: Transportation Research Board (TRB).

Kapferer, E., I. Gstach, A. Koch and C. Sedmak (eds.) 2017. *Rethinking Social Capital. Global Contributions from Theory and Practice.* Cambridge: Cambridge Scholars Publishing.

Kasraian, D. 2017. T*ransport Networks, Land Use and Travel Behaviour: A Long Term Investigation.* Dissertation, Delft University of Technology. https://doi.org/10.4233/uuid:5293031c-63c2-43bb-a53f- 750955a5c91f (15/11/2018).

Kasraian, D., K. Maat, D. Stead and B. van Wee 2016. "Long-term impacts of transport infrastructure networks on land-use change: An international review of empirical studies", in *Transport Reviews* (36) 6, 772–792.

Kauffmann, A., and M. T. W. Rosenfeld (eds.) 2012. "Städte und Regionen im Standortwettbewerb. Neue Tendenzen, Auswirkungen und Folgerungen für die Politik", in *ARL Forschungs- und Sitzungsbericht 238.* Hannover: ARL.

Kaup, G. 2013. *Ökonomie des Teilens. 15 Nutzungsgemeinschaften im Überblick.* Graz: AK Steiermark.

Kazepov, Y. (ed.) 2005. *Cities of Europe: Changing Contexts, local arrangements and the challenge to urban cohesion.* Oxford: Blackwell.

Keller, R. 2004. *Diskursforschung. Eine Einführung für SozialwissenschaftlerInnen, 2nd ed.* Wiesbaden: VS Verlag für Sozialwissenschaften.

Keller, R., A. Hirseland, W. Schneider and W. Viehöver (eds.) 2001. *Handbuch Sozialwissenschaftliche Diskursanalyse. Band 1: Theorien und Methoden.* Opladen: Leske + Budrich.

Kellerman, A. 2018. *Automated and autonomous spatial mobilities.* Cheltenham/Northampton: Edward Elgar.

Kemp, R., and D. Loorbach 2003. *Governance for Sustainability through Transition Management.* Paper for the Open Meeting of the Human Dimensions of Global Environmental Change Research Community, Montreal. https://sedac.ciesin.columbia.edu/openmtg/docs/kemp.pdf (16/3/2019).

King, A. A., and B. Baatartogtokh 2015. "How useful is the theory of disruptive innovation?", in *MIT Sloan Management Review* (57) 1, 77.

Kingdon, J. W. 1986. *Agenda, alternatives and public policy.* Boston: Little, Brown & Co.

Kitchin, R. 2015. "Making sense of smart cities: addressing present shortcomings", in *Cambridge Journal of Regions, Economy and Society* (8) 1, 131–136.

Kitchin, R., and M. Dodge 2017. "The (In)Security of Smart Cities: Vulnerabilities, Risks, Mitigation, and Prevention", in *Journal of Urban Technology* (26) 2, 1–19.

Klimaaktiv (2017). *Carsharing in Gemeinden.* www.klimaaktiv.at/mobilitaet/carsharing/Carsharing-in-Gemeinden.html (15/8/2019).

Knieling, J., and F. Othengrafen (eds.) 2009. *Planning Cultures in Europe. Decoding Cultural Phenomena in Urban and Regional Planning.* Farnham/Burlington: Ashgate.

Knoppe, M., and M. Wild (eds.) 2018. *Digitalisierung im Handel. Geschäftsmodelle, Trends und Best Practice.* Berlin: Springer.

Koch, F., S. Kabisch and K. Krellenberg 2017. "A Transformative Turn towards Sustainability in the Context of Urban-Related Studies? A Systematic Review from 1957 to 2016", in *Sustainability* (10) 1, 58.

Kollosche, I., and O. Schwedes 2016. "Mobilität im Wandel. Transformationen und Entwicklungen im Personenverkehr", in *WISO Diskurse 14/2016.* Berlin: Friedrich-Ebert-Stiftung.

Kopp, J., K. W. Axhausen and R. Gerike 2015. "Do sharing people behave differently? An empirical evaluation of the distinctive mobility patterns of free-floating car-sharing members", in *Transportation 42*, 449–469.

Kosow, H., R. Gaßner, L. Erdmann and B. Luber 2008. *Methoden der Zukunfts- und Szenarioanalyse. Überblick, Bewertung und Auswahlkriterien. Workshop report no. 103.* Bonn: Institut für Zukunftsstudien und Technologiebewertung.

Kosow, H., and C. D. León 2015. "Die Szenariotechnik als Methode der Experten- und Stakeholdereinbindung", in Niederberger/Wassermann (eds.), 217–242.

Koyama, H. 2015. *Activity Plan of Dynamic Map Study for SIP-adus. 2nd SIP-adus Workshop.* www.sip-adus.go.jp/workshop/program/speaker/profile/dm/koyama.pdf (13/10/2018).

KPMG (Klynveld, Peat, Marwick, Goerdeler) (2018). *Autonomous Vehicles Readiness Index.* https://assets.kpmg.com/content/dam/kpmg/nl/pdf/2018/sector/automotive/autonomous-vehicles-readiness-index.pdf (24/2/2019).

Kraftfahrt-Bundesamt (Federal Motor Transport Authority) (2018). *Pressemitteilung Nr. 01/2018 – Fahrzeugzulassungen im Dezember 2017 – Jahresbilanz.* www.kba.de/DE/Presse/Pressemitteilungen/2018/Fahrzeugzulassungen/pm01_2018_n_12_17_pm_komplett.html?nn=1837832 (10/12/2018).

Kucharavy, D., and R. De Guio 2011. "Application of S-shaped curves", in *Procedia Engineering 9*, 559–572.

Kutter, E. 2016. "Siedlungsstruktur und Verkehr – Zum Verständnis von Sachzwängen und individueller Verkehrserreichbarkeit in Stadtregionen", in Schwedes, O., W. Canzler and A. Knie (eds.), *Handbuch Verkehrspolitik, 2nd ed.*, 1–21. Wiesbaden: Springer.

Kuzumaki, S. 2017. *SIP Automated driving systems – Mobility bringing everyone a smile.* Keynote, SIP-adus Workshop. Tokyo. https://connectedautomateddriving.eu/wp-content/uploads/2017/02/2_Day2_PL10_Kuzumaki_final_LR.pdf (13/10/2018).

Läpple, D. 2005. "Phönix aus der Asche. Die Neuerfindung der Stadt", in Berking/Löw (eds.), 397–413.

Latour, B. 2003. "Is Re-modernization Occurring – And If So, How to Prove It? A Commentary on Ulrich Beck Theory", in *Culture & Society (20)* 2, 35–48.

Lee, T. 2018. "Waymo One, the groundbreaking self-driving taxi service, explained", in *Arstechnica 5*, December 2018. https://arstechnica.com/cars/2018/12/waymo-one-the-groundbreaking-self-driving-taxi-service-explained/ (1/12/2018).

Le Galès, P. 2002. *European Cities: Social Conflicts and Governance.* Oxford: Oxford University Press.

Leimenstoll, W. 2017. *Autonomous vehicles could have a big impact on D.C.'s budget.* Washington DC: D.C. Policy Center. www.dcpolicy-center.org/publications/autonomous-vehicles-could-have-a-big-impact-on-d-c-s-budget/ (2/2/2019).

Lemmer, K. (ed.) 2015. *Neue autoMobilität: Automatisierter Straßenverkehr der Zukunft.* Munich: acatech.

Lennert, F., and R. Schönduwe 2017. "Disrupting Mobility: Decarbonising Transport?", in Meyer/Shaheen (eds.), 213–238.

Lenz, B., and E. Fraedrich 2016. "New Mobility Concepts and Autonomous Driving: The Potential for Change", in Maurer et al. (eds.), 173–191.

Levinson, D. 2008. "Density and dispersion: The co-development of land use and rail in London", in *Journal of Economic Geography* 8, 55–77.

Levinson, D. 2015. "Climbing Mount Next: The Effects of Autonomous Vehicles on Society", in *Minnesota Journal of Law, Science & Technology* (16) 2, 787–809.

Levitas, R. 2010. "Back to the future: Wells, sociology, utopia and method", in *The Sociological Review* (58) 4, 530–547.

Li, Y., and T. Voege 2017. "Mobility as a Service (MaaS): Challenges of Implementation and Policy Required", in *Journal of Transportation Technologies* (7) 2, 95–106.

Liang, X., G. Correia and B. van Arem 2016. "Optimizing the service area and trip selection of an electric automated taxi system used for the last mile of train trips", in *Transportation Research Part E: Logistics and Transportation Review 93*, 115–129.

Libbe, J. 2018. "Smart City", in Rink/Haase (eds.), 429–449.

Lindloff, K., N. Pieper, N. C. Bandelow and D. M. Woisetschläger 2014. "Drivers of carsharing diffusion in Germany: an actor-centred approach", in *International Journal of Automotive Technology and Management* (14) 3/4, 217.

Lissandrello, E., and J. Grin 2011. "Reflexive Planning as Design and Work: Lessons from the Port of Amsterdam", in *Planning Theory & Practice* (12) 2, 223–248.

Litman, T. 2017. *Autonomous Vehicle Implementation Predictions: Implications for Transport Planning.* Victoria: Victoria Transport Policy Institute.

Loo, D. 2017. *Successful Ageing in Singapore: Urban Implications in a High-density City.* Lee Kwan Yew School of Public Policy (LKYSPP), National University of Singapore (NUS). https://lkyspp. nus.edu.sg/docs/default-source/case-studies/entry1792successful_ ageing_in_singapore_092017.pdf?sfvrsn=21d7950b_0 (13/9/2018).

Loose, W. 2010. *Aktueller Stand des Car-Sharing in Europa. Endbericht D 2.4.* Berlin: Bundesverband CarSharing.

LTA (Land Transport Authority) 2013. *Land Transport Masterplan 2013.* Singapore: LTA. www.lta.gov.sg/content/dam/ltaweb/corp/ PublicationsResearch/files/ReportNewsletter/LTMP2013Report.pdf (13/9/2018).

LTA 2018. *Smart Mobility 2030 – Singapore ITS Strategic Plan.* Singapore: LTA. www.lta.gov.sg/content/ltaweb/en/roads-and-motoring/managing-traffic-and-congestion/intelligent-transport-systems/ SmartMobility2030.html (13/9/2018).

Lund, E. 2017. *Mobility as a Service: What is it, and which problems could it solve?* https://en.trivector.se/fileadmin/user_upload/Traffic/ Whitepapers/Mobility_as_a_Service.pdf (6/3/2019).

Luo, Y., T. Chen, S. Zhang and K. Li 2015. "Intelligent Hybrid Electric Vehicle ACC with Coordinated Control of Tracking Ability, Fuel Economy, and Ride Comfort", in *IEEE Transactions on Intelligent Transportation Systems* (16) 4, 2303–2308.

MA 18 (Magistratsabteilung 18 – Stadtentwicklung und Stadtplanung Wien) 2014. *Mut zur Stadt. Stadtentwicklungsplan Wien, STEP 2025.* Vienna: Stadt Wien, MA 18.

Maertins, C. 2006. *Die Intermodalen Dienste der Bahn: Mehr Mobilität und weniger Verkehr? Wirkungen und Potenziale neuer Verkehrsdienstleistungen.* Discussion paper SP III 2006-101, Wissenschaftszentrum Berlin für Sozialforschung. Berlin: WZB.

Malodia, S., and H. Singla 2016. "A study of carpooling behaviour using a stated preference web survey in selected cities of India", in *Transportation Planning and Technology* (39) 5, 538–550.

Manderscheid, K. 2018. "From the Automobile to the Driven Subject", in *Transfers* (8) 1, 24–43. DOI: 10.3167/ TRANS.2018.080104.

Maracke, C. 2017. "Autonomes Fahren – ein Einblick in die rechtlichen Rahmenbedingungen", in *Wirtschaftsinformatik & Management 3/2017*, 62–68.

Marshall, S. 2005. *Street & Patterns.* London: Spon Press.

Marx, P. 2018. "Self-Driving cars are out. Micromobility is in". https://medium.com/s/story/self-driving-cars-will-always-be-limited-even-the-industry-leader-admits-it-c5fe5aa01699 (14/8/2019).

Maurer, M., J. C. Gerdes, B. Lenz and H. Winner (eds.) 2016. *Autonomous Driving: Technical, Legal and Social Aspects.* Berlin/ Heidelberg: Springer Vieweg.

Mayor of London 2014. *London Infrastructure Plan 2050.* Transport Supporting Paper. London: Mayor of London.

McCarthy, J., J. Bradburn, D. Williams, R. Piechocki and K. Hermans 2015. *Connected & Autonomous Vehicles: Introducing the Future of Mobility.* London: Atkins.

McShane, C. 1979. "Transforming the use of urban space: A look at the revolution in street pavements, 1880–1924", in *Journal of Urban History* (5) 3, 279–307.

McShane, C. 1994. *Down the asphalt path: The automobile and the American city.* New York, NY: Columbia University Press.

Menke, C., and J. Rebentisch (eds.) 2016. *Kreation und Depression. Freiheit im gegenwärtigen Kapitalismus.* Berlin: Kulturverlag Kadmos.

Merat, N., R. Madigan and S. Nordhoff 2017. *Human Factors, User Requirements, and User Acceptance of Ride-Sharing in Automated Vehicles.* Discussion paper 2017-Paris: International Transport Forum.

Metz, D. 2018. "Developing Policy for Urban Autonomous Vehicles: Impact on Congestion", in *Urban Science* (2) 33, 1–11.

Meyer, G., and S. Beiker (eds.) 2014. *Road Vehicle Automation.* Cham: Springer.

Meyer, G., and S. Beiker (eds.) 2016. *Road Vehicle Automation 3.* Cham: Springer.

Meyer, G., and S. Beiker (eds.) 2018. *Road Vehicle Automation 4.* Cham: Springer.

Meyer, G., and S. Beiker (eds.) 2019. *Road Vehicle Automation 5.* Cham: Springer.

Meyer, G., and S. Shaheen (eds.) 2017. *Disrupting Mobility. Impacts of Sharing Economy and Innovative Transportation on Cities.* Cham: Springer.

Michelmann, H., C. Marquardt and C. Pitzen 2017. "Autonome Busse im ÖPNV – eine zeitnahe Zukunftstechnologie?" www.thega.de/fileadmin/thega/pdf/thega-forum/2017/vortraege/session_3/ pitzen.pdf (13/9/2018).

Milakis, D., B. van Arem and B. van Wee 2017. "Policy and society related implications of automated driving: A review of literature and directions for future research", in *Journal of Intelligent Transportation Systems* (21) 4, 324–348. DOI:10.1080/15472450.2017. 1291351.

Milton Keynes Council 2011. *A Transport Vision and Strategy for Milton Keynes. Local Transport Plan 3, 2011 to 2031.* www.milton-keynes.gov.uk/assets/attach/6711/Milton_Keynes_LTP3_Appendix_A_to_F_31-03-2011.pdf (24/2/2019).

MIM (Ministerie van Infrastructuur en Milieu) 2015. *Mobiliteitsbeeld 2015.* The Hague: Kennisinstituut voor Mobiliteitsbeleid.

Ministry of Transport, Public Works and Water Management 2010. *Public transport in the Netherlands.* www.emta.com/IMG/pdf/ brochure.pdf (2/2/2019).

Minx, E., and E. Böhlke 2006. "Denken in alternativen Zukünften", in *Internationale Politik (61)* 14, 14–22.

Mitscherlich, A. 1965. *Die Unwirtlichkeit unserer Städte: Anstiftung zum Unfrieden.* Frankfurt am Main: Suhrkamp.

Mitteregger, M. 2021. "Am Ende der Straße: totale Sicherheit. Wie das Sicherheitskonzept von automatisierten und vernetzten Fahrsystemen den Straßenraum verändert", in Mitteregger et al. (eds.), AVENUE21. Planerische und politische Aspekte zu Gestaltung des Zeitalters der automatisierten Mobilität. Heidelberg: Springer Vieweg.

Mitteregger, M., A. Soteropoulos, J. Bröthaler and F. Dorner 2019. *Shared, Automated, Electric: the Fiscal Effects of the "Holy Trinity".* Proceedings of the 24th REAL CORP, International Conference on Urban Planning, Regional Development and Information Society. Karlsruhe.

MLIT (Ministry of Land, Infrastructure, Transport and Tourism) 2013. *Basic Plan on Transport Policy.* Ministry of Land, Infrastructure, Transport and Tourism. Tokyo. www.mlit.go.jp/common/001096409.pdf (13/10/2018).

Moon, S., I. Moon and K. Yi 2009. "Design, tuning, and evaluation of a full-range adaptive cruise control system with collision avoidance", in *Control Engineering Practice* (17) 4, 442–455.

Moreno, A. T. 2017. *Autonomous vehicles: Implications on an integrated land-use and transport-modelling suite.* 11th AESOP Young Academics Conference, Munich.

MOT (Ministry of Transport) 2017. *Committee on Autonomous Road Transport for Singapore.* Press release from 27/8/2017. www. mot.gov.sg/news-centre/news/Detail/Committee-on-Autonomous-Road-Transport-for-Singapore (13/9/2018).

Mumford, L. 1961. *The City in History: Its Origins, Its Transformations, and Its Prospects.* New York: Harcourt, Brace & World.

MVW (Ministerie van Verkeer en Waterstaat) 2009. *Cycling in the Netherlands.* The Hague. www.fietsberaad.nl/library/repository/ bestanden/CyclingintheNetherlands2009.pdf (2/2/2019).

Nabielek, K., D. Hamers and D. Evers 2016. *Cities in Europe: Facts and figures on cities and urban areas.* The Hague: PBL Netherlands Environmental Assessment Agency.

NACTO (National Association of City Transport Officials) 2017. *Blueprint for Autonomous Urbanism.* New York, NY. https:// nacto.org/wp-content/uploads/2017/11/BAU_Mod1_raster-sm.pdf (3/12/2018).

Navya 2017. *NAVYA We drive your future – Press Kit*. Villeurbanne. https://navya.tech/wp-content/uploads/2017/09/Navya_Press_Kit_EN_PDF_10_03_17.pdf (3/12/2018).

Neal, Z. P. 2013. *The Connected City: How Networks are Shaping the Modern Metropolis*. London/New York, NY: Routledge.

Newman, P., and J. Kenworthy 1999. *Sustainability and Cities: Overcoming Automobile Dependence*. Washington D.C.: Island Press.

Ngoduy, D. 2012. "Application of gas-kinetic theory to modelling mixed traffic of manual and ACC vehicles", in *Transportmetrica* (8) 1, 43–60.

NHTSA (National Highway Traffic Safety Administration) 2017. *Federal Automated Vehicles Policy. Accelerating the Next Revolution in Roadway Safety*. Washington DC: U.S. Department of Transportation.

Niederberger, M., and S. Wassermann (eds.) 2015. *Methoden der Experten- und Stakeholdereinbindung in der sozialwissenschaftlichen Forschung*. Wiesbaden: Springer VS.

Nielsen, J. R., H. Hovmøller, P.-L. Blyth and B. K. Sovacool 2015. "Of 'white crows' and 'cash savers': A qualitative study of travel behavior and perceptions of ridesharing in Denmark", in *Transportation Research Part A: Policy and Practice 78*, 113–123.

NLA (New London Architecture) 2017. *London's Towns: Shaping the Polycentric City*. www.guildmore.com/wp-content/uploads/2017/10/London-Towns-Publication-1-1.pdf (21/11/2018).

OECD (Organisation for Economic Co-operation and Development) 2007. *OECD Territorial Reviews*. Randstad Holland, Netherlands. Brussels: OECD.

OECD 2014. *OECD Territorial Reviews: Netherlands 2014*. OECD Publishing. DOI:10.1787/9789264209527-en.

OECD 2015. *The Metropolitan Century: Understanding Urbanisation and its Consequences*. Paris: OECD.

ÖROK (Österreichische Raumordnungs-Konferenz) 2015. *Für eine Österreichische Stadtregionspolitik. Agenda Stadtregionen in Österreich. Recommendations of the ÖREK partnership "Kooperationsplattform Stadtregion"*. Vienna: ÖREK. www.stadt-umland.at/fileadmin/sum_admin/uploads/sum_konferenzen/sum_konferenz_15/Agenda_Stadtregion_final.pdf (12/4/2019).

Ohmae, K. 1985. *Macht der Triade. Die neue Form weltweiten Wettbewerbs*. Wiesbaden: Gabler.

Ohnemus, M., and A. Perl 2016. "Shared Autonomous Vehicles: Catalyst of New Mobility for the Last Mile?", in *Built Environment* (42) 4, 589–602. DOI:10.2148/benv.42.4.589.

ORF 2018. *Wo die meisten Autos wohnen: Motorisierungsgrad in Europas Städten*. https://orf.at/v2/stories/2430040/2430061/ (6/3/2019).

Owens, J. M., R. Greene-Roesel, A. Habibovic, L. Head and A. Apricio 2018. "Reducing Conflict Between Vulnerable Road Users and Automated Vehicles", in Meyer/Beiker (eds.), 69–75.

Owyang, J., A. Samuel and A. Grenville 2014. *Sharing is the New Buying: How to Win in the Collaborative Economy*. Vancouver: Vision Critical/Crowd Companies.

Pangbourne, K., M. Mladenovic, D. Stead and D. Milakis 2019. "Questioning Mobility as a Service: Unanticipated societal and governance implications", in *Transportation Research Part A: Policy and Practice*.

Parkin, J., B. Clark, W. Clayton, M. Ricci and G. Parkhurst 2016. *Understanding interactions between autonomous vehicles and other road users: A Literature Review. Project report*. Bristol: University of the West of England.

Paskaleva, K. A. 2011. "The smart city: A nexus for open innovation?", in *Intelligent Buildings International (3)* 3, 153–171. DOI:10.1080/17508975.2011.586672.

Paterson, M. 2007. *Automobile Politics: Ecology and Cultural Political Economy*. Cambridge: Cambridge University Press.

Peer, C. 2016. "Stadtalltag als Labor. Forschungsperspektiven zur Koexistenz internationaler Verflechtungen und lokaler Wissenskulturen im Rahmen von Living Labs", in Staubmann (ed.), 315–329.

Perret, F., F. Bruns, L. Raymann, S. Hofmann, R. Fischer, C. Abegg, P. de Haan, R. Straumann, S. Heuel, M. Deublein and C. Willi 2017. *Einsatz automatisierter Fahrzeuge im Alltag – Denkbare Anwendungen und Effekte in der Schweiz*. Zurich: EBP. Basler Fonds.

Petermann, T. (ed.) 1903. *Die Großstadt. Vorträge und Aufsätze zur Städteausstellung. Jahrbuch der Gehe-Stiftung*. Dresden: Gehe-Stiftung.

Picon, A. 2015. "The limits of intelligence: On the challenges faced by smart cities", in *Geographies 7*, 77–83.

Plötz, P., U. Schneider, J. Globisch and E. Dütschke 2014. "Who will buy electric vehicles? Identifying early adopters in Germany", in *Transportation Research Part A: Policy and Practice 67*, 96–109.

POLIS (European Cities and Regions Networking for Innovative Transport Solutions) 2018. *Road Vehicle Automation and Cities and Regions*. Brussels.

Porter, M., and J. Heppelmann 2014. "How Smart, Connected Products Are Transforming Competition", in *Harvard Business Review* (92) 11, 65–88.

PostAuto Schweiz AG 2016. "Autonomer Bus 'SmartShuttle' in Sitten". www.postauto.ch/de/file/19206/download?token=Uv6GobkC (6/3/2019).

Priddat, B. P. 2015. "Share Economy: mehr Markt als Gemeinschaft", in *Wirtschaftsdienst: Zeitschrift für Wirtschaftspolitik* (95) 2, 98–101.

Prime Minister's Cabinet 2017. *Public-Private ITS Initiative/Roadmaps 2017: Toward implementation of various highly automated driving systems in society*. Presentation at Strategic Conference for the Advancement of Utilizing Public and Private Sector Data, Strategic Headquarters for the Advanced Information and Telecommunications Network Society. https://japan.kantei.go.jp/policy/it/itsinitiative_roadmap2017.pdf (13/10/2018).

Provincie Noord-Holland 2015. "Structuurvisie Noord-Holland 2040. Kwaliteit door Veelzidigheid". www.noord-holland.nl/Onderwerpen/Ruimtelijke_inrichting/Structuurvisie_en_PRV/Beleidsdocumenten/Structuurvisie_Noord_Holland_2040.org (24/2/2019).

PwC (PricewaterhouseCoopers GmbH Wirtschaftsprüfungsgesellschaft) 2018. *Five trends transforming the Automotive Industry*. www.pwc.at/de/publikationen/branchen-und-wirtschaftsstudien/eascy-five-trends-transforming-the-automotive-industry_2018.pdf. (6/3/2019).

Raimondi, F. M., and M. Melluso 2008. "Fuzzy motion control strategy for cooperation of multiple automated vehicles with passengers comfort", in *Automatica* (44) 11, 2804–2816.

Rammler, S. 2016. *Digitaler Treibstoff. Chancen und Risiken des Einsatzes digitaler Technologien und Medien im Mobilitätssektor*. Study series of the Hans Böckler Foundation. www.boeckler.de/pdf/p_study_hbs_310.pdf (2/2/2019).

Rammler, S., and M. Weider (eds.) 2011. *Das Elektroauto. Bilder für eine zukünftige Mobilität*. Berlin/Münster: LIT.

Rauws, W. 2017. "Embracing Uncertainty Without Abandoning Planning: Exploring an Adaptive Planning Approach for Guiding Urban Transformations", in *DisP – The Planning Review* (53) 1, 32–45.

Rehrl, K., and C. Zankl 2018. *Digibus 2017: Erfahrungen mit dem ersten selbstfahrenden Shuttlebus auf öffentlichen Straßen in Österreich*. Salzburg: Salzburg Research.

Reichow, H. B. 1959. *Die autogerechte Stadt – Ein Weg aus dem Verkehrs-Chaos*. Ravensburg: Otto Maier.

Reid C. 2017. *How the Dutch Really Got Their Cycleways: Bike Boom*. Washington, D.C.: Island Press.

Riegler, D. 2018. "'Micro Mobility': Die E-Scooter in Wien", *ORF*, 17/11/2018, https://fm4.orf.at/stories/2947864/ (11/3/2019).

Riegler, S., M. Juschten, R. Hössinger, R. Gerike, L. Rößger, B. Schlag, W. Manz, C. Rentschler and J. Kopp 2016. *CarSharing 2025 – Nische oder Mainstream?* Munich: Ifmo (Institut für Mobilitätsforschung). www.ifmo.de/files/publications_content/2016/ifmo_2016_Carsharing_2025_de.pdf (6/3/2019).

Rietdorf, W. (ed.) 2001. *Auslaufmodell Europäische Stadt. Neue Herausforderungen und Fragestellungen am Beginn des 21. Jahrhunderts*. Berlin: Verlag für Wissenschaft und Forschung.

Riggs, W., N. Larco, G. Tierney, M. Ruhl, J. Karlin-Resnick and C. Rodier 2019. "Autonomous vehicles and the built environment: exploring the impacts on different urban contexts", in Meyer/Beiker (eds.), 221–232.

Rink, D., and A. Haase (eds.) 2018. *Handbuch Stadtkonzepte. Analysen, Diagnosen, Kritiken und Visionen*. Stuttgart: utb.

Ritt, T. 2016. *Smart City – Zukunftskonzept oder Marketing mit Nebenwirkungen?* https://wien.arbeiterkammer.at/interessenvertretung/meinestadt/sozialestadt/Thomas_Ritt_17.2.2016.pdf (2/2/2019).

Ritz, J. 2018. *Mobilitätswende – autonome Autos erobern unsere Straßen. Ressourcenverbrauch, Ökonomie und Sicherheit.* Wiesbaden: Springer.

Robinson, J. 2006. "Global and World Cities: A View from the Map", in Brenner/Keil (eds.), 217–223.

Robinson, J., S. Burch, S. Talwar, M. O'Shea and M. Walsh 2011. "Envisioning sustainability: recent progress in the use of participatory backcasting approaches for sustainability research", in *Technological Forecasting and Social Change* (78) 5, 756–768.

Rodenstein, M. 2018. "Rekonstruktion und soziales Gedächtnis – Wie Erinnerungen unsere Städte verändern", in Gestring/Wehrheim (eds.), 237–255.

Rodriguez P., J. P. Nuñez Velasco, H. Farah, and M. Hagenzieker 2016. *Safety of pedestrians and cyclists when interacting with self-driving vehicles: A case study of the WEpods.* ITRL Conference on Integrated Transport 2016: Connected & Automated Transport Systems, Stockholm.

Rogers, E. M. 2003. *Diffusion of innovations.* New York, NY: Free Press.

Roland Berger 2016. Internationale Best-Practice-Studie "Intelligente Vernetzung. Innovative und beispielhafte IKT-Projekte aus den Anwendungssektoren Bildung, Energie, Gesundheit, Verkehr und Verwaltung." www.rolandberger.com/publications/publication_pdf/roland_berger_in_best_practice_studie_iiv_1.pdf (2/2/2019).

Roorda, C., J. Wittmayer, P. Henneman, F. van Steenbergen, N. Frantzeskaki and D. Loorbach 2014. *Transition Management in the Urban Context: Guidance Manual.* Rotterdam: DRIFT, Erasmus University.

Rosa, H. 2009. *High-Speed Society: Social Acceleration, Power, and Modernity.* Pennsylvania State University Press.

Rosa, H. 2015. *Social Acceleration: A New Theory of Modernity.* Trans. Jonathan Trejo-Mathys. New York, NY: Columbia University Press.

Rotmans, J., R. Kemp and M. van Asselt 2001. "More evolution than revolution: Transition management in public policy", in *Foresight* (3) 1. DOI:10.1108/14636680110803003.

Rückert-John, J. (ed.) 2013. *Soziale Innovation und Nachhaltigkeit. Perspektiven sozialen Wandels.* Wiesbaden: VS Verlag für Sozialwissenschaften.

Rupprecht, S., S. Buckley, P. Crist, and J. Lappin 2018. "'AV-Ready' Cities or 'City-Ready' AVs?", in Meyer/Beiker (eds.), 223–233. DOI:10.1007/ 978-3-319-60934-8_18.

SAE International 2018. *Taxonomy and Definitions for Terms Related to Driving Automation Systems for On-Road Motor Vehicles – J3016,* June 2018. www.sae.org/standards/content/j3016_201806/ (2/2/2019).

Sänn, A., S. Richter and C. K. Fraunholz 2017. "Car-to-X als Basis organisationaler Transformation und neuer Mobilitätsdienstleistungen", in *Wirtschaftsinformatik & Management 5/2017,* 60–71.

Safdie, M., and W. Kohn 1998. *The City after the Automobile: An Architect's Vision.* Colorado : Westview Press.

SAFER 2017. *Co-Creation Lab: How can autonomous transport systems bring value in cities?* SAFER – Vehicle and Traffic Safety Centre at Chalmers, Gothenburg. www.saferresearch.com/about (19/11/2018).

Salonen, A. O., and N. Haavisto 2019. "Towards Autonomous Transportation: Passengers' Experiences, Perceptions and Feelings in a Driverless Shuttle Bus in Finland", in *Sustainability* (11) 588, 1–19.

Sammer, G., T. Uhlmann, W. Unbehaun, A. Millonig, B. Mandl, J. S. Dangschat and R. Mayr 2013. "Identification of mobility-impaired persons and analysis of their travel behavior and needs", in *Journal of Transportation Research Board 2320,* 46–54.

Saunders, P. 1986. *Social Theory and the Urban Question.* London: Hutchinson Education.

Schaefer, K.E. and E. R Straub 2016. *Will Passengers Trust Driverless Vehicles? Removing the steering wheel and pedals.* IEEE International Multidisciplinary Conference on Cognitive Methods in Situation Awareness and Decision Support Proceedings (CogSIMA 2015), San Diego, CA.

Schäpke, N., F. Stelzer, M. Bergmann, M. Singer-Brodowski, M. Wanner, G. Caniglia and D. J. Lang 2017. *Reallabore im Kontext transformativer Forschung. Ansatzpunkte zur Konzeption und Einbettung in den internationalen Forschungsstand.* IETSR Discussion Papers in Transdisciplinary Sustainability Research1/2017. Lüneburg: Leuphana-Universität, Institut für Ethik und Transdisziplinäre Nachhaltigkeitsforschung.

Scheiner, J. 2005. "Wohnen und Verkehr. Zusammenhänge zwischen Wohnmobilität, Wohnsituation, Standortbewertung und Verkehrsverhalten", in *Arbeitspapiere des Fachgebiets Verkehrswesen und Verkehrsplanung 14.* Dortmund: Universität Dortmund.

Scheiner, J. 2009. *Sozialer Wandel, Raum und Mobilität. Empirische Untersuchungen zur Subjektivierung der Verkehrsnachfrage.* Wiesbaden: VS Verlag für Sozialwissenschaften.

Scheltes, A. 2018. "Zelfrijdende voertuigen: kans voor het ov in Nederland?", in *NM Magazine.* www.nm-magazine.nl/artikelen/zelfrijden-de-voertuigen-kans-voor-het-ov-in-nederland/ (24/2/2019).

Scheuvens, R., S. Groh, D. Allmeier and M. Weisböck 2016. Wien: *polyzentral. Forschungsstudie zur Zentrenentwicklung Wiens.* Workshop report 158. Vienna: MA 18.

Schmitz, S. 2001. *Revolutionen der Erreichbarkeit. Gesellschaft, Raum und Verkehr im Wandel, Stadtforschung aktuell, vol. 83.* Wiesbaden: VS Verlag für Sozialwissenschaften.

Schneidewind, U. 2010. "Ein institutionelles Reformprogramm zur Förderung transdisziplinärer Nachhaltigkeitsforschung", in *GAIA: Ecological Perspectives for Science and Society* (19) 2, 122–128.

Schneidewind, U., and H. Scheck 2013. "Die Stadt als 'Reallabor' für Systeminnovationen", in Rückert-John (ed.), 229–248.

Schneidewind, U., and M. Singer-Brodowski 2015. "Vom experimentellen Lernen zum transformativen Experimentieren: Reallabore als Katalysator für eine lernende Gesellschaft auf dem Weg zu einer Nachhaltigen Entwicklung", in *Zeitschrift für Wirtschafts- und Unternehmensethik* (16) 1, 10–23.

Schoitsch, E., C. Schmittner, Z. Ma and T. Gruber 2016. "The need for safety and cyber-security co-engineering and standardization for highly automated automotive vehicles", in Schulze et al. (eds.), 251– 261).

Scholl, G., M. Gossen, M. Grubbe and T. Brumbauer 2013. *Alternative Nutzungskonzepte – Sharing, Leasing und Wiederverwendung. Vertiefungsanalyse 1. PolRess AP2 – Politikansätze und -instrumente.* Berlin: Institut für ökologische Wirtschaftsforschung (iöw).

Schreier, H., U. Becker and J. Heller 2015. *Endbericht Evaluation CarSharing (EVA-CS),* Landeshauptstadt München. www.ris-muenchen.de/RII/RII/DOK/SITZUNGSVORLAGE/3885730.pdf (14/8/2019).

Schreier, H., C. Grimm, U. Kurz, B. Schwieger, S. Keßler and G. Möser 2018. *Analyse der Auswirkungen des Car-Sharing in Bremen. Endbericht.* https://senatspressestelle.bremen.de/sixcms/media.php/13/20180507_Endbericht_Bremen.pdf (14/8/2019).

Schrepfer, J., J. Mathes, V. Picron and H. Bath 2018. "Automatisiertes Fahren und seine Sensorik im Test", in *ATZ – Automobiltechnische Zeitschrift 1/2018,* 29–37.

Schubert, K. 1985. "Wien", in Friedrichs (ed.), 347–574.

Schulz, D., and S. Gilbert 1996. "Women and Transit Security: A New Look at an Old Issue", in *Women's Travel Issues,* Second National Conference. Washington DC: Federal Highway Administration.

Schulz-Montag, B., and M. Müller-Stoffels 2006. "Szenarien – Instrumente für Innovations- und Strategieprozesse", in Wilms (ed.), 381– 397.

Schulze, T., B. Müller and G. Meyer (eds.) 2016. *Advanced Microsystems for Automotive Applications 2016: Smart Systems for the Automobile of the Future.* Cham: Springer.

Schwarz, U. 2014. "Reflexive Moderne und Architektur Revisited", in Buchert (ed.), 188–211.

Schwedes, O. 2017. *Verkehr im Kapitalismus.* Münster: Westfälisches Dampfboot.

Schwedes, O., W. Canzler and A. Knie (eds.) 2016. *Handbuch Verkehrspolitik. 2nd ed.* Wiesbaden: VS Verlag für Sozialwissenschaften.

Schwedes, O., and S. Rammler 2012. *Mobile Cities – Dynamiken weltweiter Stadt- und Verkehrsentwicklung, 2nd ed.* Berlin: LIT.

Seider, C., and P. Schmitz 2017. *Security-Prognosen 2018: Cyber-Sicherheit für das vernetzte Auto.* www.security-insider.de/cyber-sicherheit-fuer-das-vernetzte-auto-a-671961/ (24/2/2019).

Senatsverwaltung für Bau- und Wohnungswesen (ed.) 1990. *Stadterneuerung Berlin.* Berlin.

Sennett, R. 1986. *The Fall of Public Man.* London: Faber.

Sennett, R. 2019. *Building and Dwelling: Ethics for the City.* London: Penguin.

Seppelt, B. D., and J. D. Lee 2007. "Making adaptive cruise control (ACC) limits visible", in *International Journal of Human Computer Studies* (65) 3, 192–205.

SFMTA (San Francisco Municipal Transportation Agency) 2016a. *City of San Francisco: Meeting the Smart City Challenge. Transportation Vision of City of San Francisco.* https://cms.dot.gov/sites/dot.gov/files/docs/San%20Francisco%20Vision%20Narrative.pdf (10/11/2018).

SFMTA 2016b. *City of San Francisco: Meeting the Smart City Challenge. Vol. 1.* San Francisco. https://cms.dot.gov/sites/dot.gov/files/docs/San-Francisco-SCC-Technical-Application.pdf (10/11/2018).

SFMTA 2018. *SFMTA Strategic Plan.* www.sfmta.com/sites/default/files/reports-and-documents/2018/04/sfmta_strategic_plan.pdf (10/11/2018).

Shaheen, S., and N. Chan 2016. "Mobility and the Sharing Economy: Potential to Facilitate the First- and Last-Mile Public Transit Connections", in *Built Environment* (42) 4, 573–588.

Shladover, S. E. 2018. *Practical Challenges to Deploying Highly Automated Vehicles.* Presentation at Drive Sweden. Göteborg.

Siebel, W. 2015. *Die Kultur der Stadt.* Frankfurt am Main: Suhrkamp.

Siebel, W. (ed.) 2004. *Die europäische Stadt.* Frankfurt am Main: Suhrkamp.

Siems, F. U., and M. C. Papen (eds.) 2018. *Kommunikation und Technik. Ausgewählte neue Ansätze im Rahmen einer interdisziplinären Betrachtung, vol. 28.* Wiesbaden: Springer.

Simmel, G. 1964. "The Metropolis and Mental Life", in: K. H. Wolff (ed.), *The Sociology of Georg Simmel,* 409–420.

Simon, D. 2006. "The World City Hypothesis: Reflections from the Periphery", in Brenner/Keil (eds.), 203–209.

Singer-Brodowski, M., R. Beecroft and O. Parodi 2018. "Learning in Real-World Laboratories: A Systematic Impulse for Discussion", in *GAIA: Ecological Perspectives for Science and Society* (27) S1, 23–27.

Sinning, H. 2008. "Urban Governance und Stadtentwicklung. Zur Rolle des Bürgers als aktiver Mitgestalter und Ko-Produzent", in *Newsletter Wegweiser Bürgergesellschaft 12/2008,* 1–10.

SIP (Cross-ministerial Strategic Innovation Promotion Program) 2017. *What is the Cross-ministerial Strategic Innovation Promotion Program?* www8.cao.go.jp/cstp/panhu/sip_english/4-6.pdf (13/10/2018).

SIP-ADUS 2016. A Revolutionary Traffic System for Citizens and Cities – Freedom of Movement and Safety for All. www8.cao.go.jp/cstp/panhu/sip_english/30-33.pdf (22/6/2019).

Sivak, M., and B. Schoette 2015. *Motion Sickness in Self-Driving Vehicles.* Ann Arbor, MI: University of Michigan. Transportation Research Institute.

Smart Nation Singapore 2018. *A Singapore Government that is Digital to the Core and Serves with Heart.* Press release of "Launch of Digital Government Blueprint" by Smart Nation Singapore from 5/6/2018. www.smartnation.sg/newsroom/press-re-leases/launch-of-digital-government-blueprint--a-singapore-government-that-is-digital-to-the-core--and-serves-with-heart (13/9/2018).

Smith, G., J. Sochor and M. A. Karlsson 2017. *Mobility as a Service: Implications for future mainstream public transport.* 15th International Conference Series on Competition and Ownership in Land Passenger Transport (Thredbo), Stockholm.

Smith, S. 2015. *British "New City" Chooses Driverless Cars Over Monorail.* https://nextcity.org/daily/entry/podcars-milton-keynes-la-freeway-tunnel-jeddah-metro-star-architect (2/2/2019).

Sochor, J., H. Arby, M. A. Karlsson and S. Sarasini 2017. *A topological approach to Mobility as a Service: A proposed tool for understanding requirements and effects, and for aiding the integration of societal goals.* ICoMaaS 2017 Proceedings, 187–208. www.tut.fi/verne/aineisto/S6_Sochor.pdf (13/7/2018).

Sommer, C. 2016. *Mobilitäts- und Angebotsstrategien in ländlichen Räumen. Planungsleitfaden für Handlungsmöglichkeiten von ÖPNV-Aufgabenträgern und Verkehrsunternehmen unter besonderer Berücksichtigung wirtschaftlicher Aspekte flexibler Bedienungsformen.* Berlin: BMVI.

Sonnberger, M., and M. Gross 2018. "Rebound Effects in Practice: An Invitation to Consider Rebound from a Practice Theory Perspective", in *Ecological Economics 154,* 14–21.

Soteropoulos, A., M. Berger and F. Ciari 2018a. "Impacts of automated vehicles in travel behaviour and land use: An international review of modelling studies", in *Transport Reviews* (39) 1, 29–49.

Soteropoulos, A., M. Mitteregger and J. Bröthaler 2018b. "Der Individualverkehr der Zukunft: Fiskalische Effekte von Automatisierung, Vernetzung und Elektrifizierung", in Suitner et al. (eds.), 97–112.

Soteropoulos, A., M. Mitteregger, M. Berger and J. Zwirchmayr 2020. "Automated drivability: Toward an assessment of the spatial deployment of level 4 automated vehicles", in *Transportation Research Part A: Policy and Practice 136,* 64–84.

Stadt Wien 2006. "Masterplan Verkehr Wien 2003 – Kurzfassung. Aktualisierter Nachdruck". Vienna: MA 18. www.eltis.org/sites/default/files/case-studies/documents/masterplan_verkehr1_3.pdf (14/9/2019).

Stadt Wien (2017). "Projektplanung – U-Bahn-Ausbau U2 und U5". www.wien.gv.at/stadtentwicklung/projekte/verkehrsplanung/u-bahn/ u2u5/projektplanung.html (18/4/2017).

Stark, E., D. T Egnor, R. Cash and K. Patterson 2019. "U.S. Patent Application No. 15/716,872".

Statistics Austria 2018. *Kfz-Bestand 1960–2017.* Vienna. www.statistik.at/web_de/statistiken/energie_umwelt_innovation_mobilitaet/verkehr/strasse/kraftfahrzeuge_-_bestand/index.html (6/8/2018).

Statistics Austria 2019a. *Gebarungsstatistik der Länder und Gemeinden 2007–2017.* Vienna. http://statistik.at/web_de/statistiken/wirtschaft/oeffentliche_finanzen_und_steuern/oeffentliche_finanzen/gebarungen_der_oeffentlichen_rechtstraeger/index.html (15/8/2018).

Statistics Austria (2019b). *Kraftfahrzeuge – Neuzulassungen.* Vienna. www.statistik.at/web_de/statistiken/energie_umwelt_innovation_mobilitaet/verkehr/strasse/kraftfahrzeuge_-_neuzulassungen/index.html (15/8/2019).

Stead, D., and E. Meijers 2015. *Urban planning and transport infrastructure provision in the Randstad Netherlands – A Global City Cluster.* www.itf-oecd.org/sites/default/files/docs/stead.pdf (2/2/2019).

Steding, D., A. Herrmann and M. Lange 2004. *Carsharing – sozialinnovativ und kulturell selektiv? Möglichkeiten und Grenzen einer nachhaltigen Mobilität.* Zentrum für Umweltforschung der Westfälischen Wilhelms-Universität Münster, UFO-Berichte, vol. 5. Münster: WWUM.

Staubmann, H. (ed.) 2016. *Soziologie in Österreich – Internationale Verflechtungen.* Innsbruck: innsbruck university press.

Steenbergen, B. van (ed.) 1994. *The Condition of Citizenship.* London/Thousand Oaks, CA/New Delhi: Sage.

Stirling, A. 2006. "Precaution, Foresight and Sustainability: Reflection and Reflexivity in the Governance of Science and Technology", in Voß et al. (eds.), 225–272.

Stirling, A. 2014. *Emancipating Transformations: From Controlling "the Transition" to Culturing Plural Radical Progress.* Brighton: STEPS Centre.

STRIA 2019. *Roadmap on Connected and Automated Transport: Road, Rail and Waterborne.* Brussels: European Commission.

Stüber, J. 2018. *EU erlaubt Fusion der Carsharing-Anbieter Car2Go und DriveNow.* www.gruenderszene.de/automotive-mobility/carsharing-fusion-eu?interstitial (17/7/2019).

Suitner, J., J. S. Dangschat and R. Giffinger (eds.) 2018. *Die digitale Transformation von Stadt, Raum und Gesellschaft,* Jahrbuch des Departments für Raumplanung der TU Wien 2018, vol. 6. Vienna/Graz: Neuer Wissenschaftlicher Verlag.

Tettamanti, T., I. Varga and Z. Szalay 2016. "Impacts of Autonomous Cars from a Traffic Engineering Perspective", in *Periodica Polytechnica Transportation Engineering* (44) 4, 244–250.

TfL (Transport for London) 2017. *Healthy Streets for London: Prioritising walking, cycling and public transport to create a healthy city.* Transport for London & The Mayor of London. http://content.tfl.gov.uk/healthy-streets-for-london.pdf (21/11/2018).

Thakur, P., R. Kinghorn and R. Grace 2016. *Urban form and function in the autonomous era*. 38th Australasian Transport Research Forum ATRF 2016. Melbourne.

The Economist 2018. *"Missing the bus: Public transport is in decline in many wealthy cities"*, 21/6/2018, www.economist.com/international/2018/06/21/public-transport-is-in-decline-in-many-wealthy-cities (20/8/2018).

The Urban Task Force 2003. *Towards an urban renaissance. Final Report of the Urban Task Force*. London: Routledge.

Thompson, R. L., C. Higgins and J. M. Howell 1991. "Personal Computing: Towards a Conceptual Model of Utilization", in *MIS Quarterly* (15) 1, 125–143.

Tokyo Metropolitan Government 2016. *New Tokyo. New Tomorrow – The Action Plan for 2020*. www.metro.tokyo.jp/english/about/plan/documents/pocket_english.pdf (13/10/2018).

UN (United Nations) 2008. *An Overview of Urbanization, Internal Migration, Population Distribution and Development in the World*. New York.

UN 2018. *Revision of World Urbanization Prospects*. https://population.un.org/wup/Publications/Files/WUP2018-KeyFacts.pdf (28/2/2019).

UNECE (United Nations Economic Commission for Europe) 2012. *From Transition to Transformation, Sustainable and Inclusive Development in Europe and Asia*. New York, NY: United Nations.

URA (Singapore Urban Research Authority) 2016. "A Look at Self-Driving Cars. Imagining our Future: A look at shifting landscapes and new ideas", in *Skyline 03*, 31–32. Singapore: Urban Redevelopment Authority (URA).

Urry, J. 2004. "The 'System' of Automobility", in *Theory, Culture & Society* (21) 4–5, 25–39.

US Senate Hearing 2018. *The Future of Autonomous Vehicles in America*, 13/6/2018. Video. www.youtube.com/watch?v=2xxpPYsreIM (2/11/2019).

UTC (Urban Transport Committee) 2014. *Gothenburg 2035 Transport Strategy for a Close-Knit City*. City of Gothenburg. https://goteborg.se/wps/wcm/connect/6c603463-f0b8-4fc9-9cd4-c1e934b41969/Trafikstrategi_eng_140821_web.pdf?MOD=AJPERES (19/11/2018).

VCÖ (Verkehrsclub Österreich) 2011. *Wie hohe Lebensqualität in Städten durch Elektro-Mobilität entsteht*. www.alexanderstiasny.at/include/download/files/29_2011-004vcoe-factsheet-urbane_e-mobilitaet.pdf (6/3/2019).

VCÖ 2017. *Jede dritte Autofahrt kürzer als 5 Kilometer – Hoher Spritverbrauch bei kurzen Autofahrten*. www.vcoe.at/presse/presse-aussendungen/detail/vcoe-wiener-autos-vor-allem-fuer-kurze-strecken-im-einsatz (6/3/2019).

VCÖ 2018a. *Mehr als 100.000 Carsharing-Haushalte in Österreich – Potenzial um ein Vielfaches höher*. www.vcoe.at/presse/presseaussendungen/detail/carsharing-haushalte-potential-2018 (6/3/2019).

VCÖ 2018b. *Großes Potenzial für Sharing und neue Mobilitätsservices*. www.vcoe.at/publikationen/vcoe-factsheets/detail/vcoe-factsheet-2018-10-sharing-und-neue-mobilitaetsloesungen (6/3/2019).

VDA – Verband der Automobilindustrie 2015. "Automatisierung. Von Fahrerassistenzsystemen zum automatisierten Fahren", in *VDA Magazin*, September 2015. Berlin.

Venkatesh, V., and F. D. Davis 2000. "A theoretical extension of the technology acceptance model: Four longitudinal field studies", in *Management Science* (46) 2, 186–204.

Venkatesh, V., M. G. Morris, G. B. Davis and F. D. Davis 2003. "User acceptance of information technology" Toward a unified view", in *Management Information Systems Quarterly* (27) 3, 425–478.

Venkatesh, V., J. Y. L. Thong and X. Xu 2016. "Unified Theory of Acceptance and Use of Technology: A Synthesis and the Road Ahead", in *Journal of the Association for Information Systems* (17) 5, 328–376.

Viehöver, W. 2001. "Diskurse als Narrationen", in Keller et al. (eds.), 177–206.

Vogel, M., R. Hamon, G. Lozenguez, L. Merchez, P. Abry, J. Barnier, P. Borgnat, P. Flandrin, I. Mallon and C. Robardet 2014. "From bicycle sharing system movements to users: a typology of Vélo cyclists in Lyon based on large-scale behavioural dataset", in *Journal of Transport Geography 41*, 280–291

Vogler, T., J. Labus and O. Specht 2018. "Mögliche Auswirkungen von Digitalisierung auf die Organisation von Handelsunternehmen", in Knoppe/Wild (eds.), 149–172.

Vosniadou, S., and A. Ortony (eds.) 1989. *Similarity and analogical reasoning*. Cambridge: Cambridge University Press.

Voß, J.-P., and R. Kemp 2006. "Sustainability and reflexive governance: introduction", in Voß et al. (eds.), 3–26.

Voß, J.-P., D. Bauknecht and R. Kemp (eds.) 2006. *Reflexive Governance for Sustainable Development*. Cheltenham & Northampton: Edward Elgar.

Wachenfeld, W., H. Winner, C. Gerdes, B. Lenz, M. Maurer, S. A. Beiker, E. Fraedrich and T. Winkle 2016. "Use Cases for Autonomous Driving", in Maurer et al. (eds.), 9–37.

Wadud, Z., D. MacKenzie and P. Leiby 2016. "Help or hindrance? The travel, energy and carbon impacts of highly automated vehicles", in *Transport Research, Part A: Policy and Practice* (86) 4, 1–18.

Wageningen University & Research 2016. *Minister Schultz van Haegen launches test phase of WEpod self-driving vehicles*. Press release from 28/6/2016. www.wur.nl/en/news-article/Minister-Schultz-van-Haegen-launches-test-phase-of-WEpod-selfdriving-vehicles.htm (24/2/2019).

Wagner, H., and S. Kabel 2018. "Digitalisierung – Motor für innovative Geschäftsmodelle im Umfeld des hochautomatisierten Fahrens", in: Fend/Hofmann (eds.), Chap. 18 .

Watanabe, H. 2016. *A Revolutionary Traffic System for Citizens and Cities – Freedom of Movement and Safety for All*. SIP-adus 30–33. www8.cao.go.jp/cstp/panhu/sip_english/30-33.pdf (13/10/2018).

WBGU (German Advisory Council on Global Change) 2011. *World in Transition: A Social Contract for Sustainability*. Berlin: WBGU. www.wbgu.de/fileadmin/user_upload/wbgu/publikationen/hauptgutachten/hg2011/pdf/wbgu_jg2011_en.pdf (19/3/2021).

WBGU 2016. *Humanity on the Move: Unlocking the Transformative Power of Cities*. Flagship Report. Berlin: WBGU. www.wbgu.de/fileadmin/user_upload/wbgu/publikationen/hauptgutachten/hg2016/pdf/hg2016_en.pdf (19/3/2021) .

Weber, M. 1958. *The City*. Trans. G. Neuwirth. New York, NY: Free Press.

Wefering, F., S. Rupprecht, S. Bührmann and S. Böhler-Baedeke 2014. *Developing and Implementing a Sustainable Urban Mobility Plan. Guidelines*. Brussels: European Commission.

Wegener, M. 2009. "Modelle der räumlichen Stadtentwicklung – alte und neue Herausforderungen". Presentation at the 10th Aachen Colloquium "Mobilität und Stadt: Ein Blick zurück – ein Blick voraus". In *Stadt Region Land Bericht 87*, 73–81. Aachen: RWTH Aachen, Institut für Stadtbauwesen und Stadtverkehr.

Wegener, M., and F. Fürst 1999. *Land-Use Transport Interaction – State of the Art. Berichte aus dem Institut für Raumplanung 46*. Dortmund: Technische Universität Dortmund.

Wehrhahn, R. 2016. "Bevölkerung und Migration", in Freytag et al. (eds.), 39–66.

Weikl, S., and K. Bogenberger 2013. "Relocation Strategies and Algorithms for Free-Floating Car Sharing Systems", in *IEEE Intelligent Transportation Systems Magazine*, 100–111.

Weisbecker, A., M. Burmester and A. Schmidt (eds.) 2015. *Mensch und Computer 2015 – Workshopband*. Stuttgart: DeGruyter.

Welch, D., and E. Behrmann 2018. *Who's Winning the Self-Driving Car Race?* New York, NY: Blomberg. www.bloomberg.com/news/features/2018-05-07/who-s-winning-the-self-driving-car-race (24/2/2019).

White, J. M. 2016. "Anticipatory logics of the smart city's global imaginary", in *Urban Geography* (37) 4, 572–589.

Willis, K. S., and A. Aurigi 2018. *Digital and Smart Cities*. Abdingdon/New York, NY: Routledge.

Wilms, F. E. 2006. "Szenarien sind Systeme", in Wilms (ed.), 39–60.

Wilms, F. E. P. (ed.) 2006. *Szenariotechnik. Vom Umgang mit der Zukunft*. Bern/Stuttgart/Vienna: Haupt.

Wiseman, Y. 2017. "Self-Driving Car: A Computer Will Park for You", in *International Journal of Engineering & Technology for Automobile Security* (1) 1, 9–16.

Wulfhorst, G. 2003. *Flächennutzung und Verkehrsverknüpfung an Personenbahnhöfen – Wirkungsabschätzung mit systemdynamischen Modellen.* Stadt, Region, Land, Bericht 49. Aachen: RWTH Aachen, Institut für Stadtbauwesen und Stadtverkehr.

Yap, M. D., G. Correia and B. van Arem 2016. "Preferences of travellers for using automated vehicles as last mile public transport of multimodal train trips", in *Transportation Research Part A: Policy and Practice* 94, 1–16.

Zankl, C., and K. Rehrl 2017. "Digibus 2017 – Erfahrungen mit dem ersten selbstfahrenden Shuttlebus auf öffentlichen Straßen in Österreich". Salzburg: Salzburch Research.

Zhang, W., S. Guhathakurta and J. Fang 2015. "Exploring the Impact of Shared Autonomous Vehicles on Urban Parking Demand: An Agent-based Simulation Approach", in *Sustainable Cities and Society 19*, 34–45.

Ziegler, A. 2012. "Individual characteristics and stated preferences for alternative energy sources and propulsion technologies in vehicles: A discrete choice analysis for Germany", in *Transportation Research Part A: Policy and Practice* (46) 8, 1372–1385.

Zweck, A., D. Holtmannspötter, M. Braun, M. Hirt, S. Kimpeler and P. Warnke 2015. *Gesellschaftliche Veränderungen 2030. Ergebnisband 1 zur Suchphase von BMBF-Foresight Zyklus II.* Dusseldorf: Innovationsbegleitung und Innovationsberatung der VDI Technologiezentrum GmbH.

Printed in the United States
by Baker & Taylor Publisher Services